T0203721

Endorsements

"Moseley is no stranger to the discreet charms of the press corps."
The Washington Post

"If I'm ever busted for something I never did, I want Matt Moseley's number sewn into my underwear."
Sam Kashner
Author, Contributing Editor, *Vanity Fair*

"Matthew L. Moseley's Ignition is a smart and witty manifesto that skillfully illuminates the art of winning over public sentiment. Every page crackles with hard-earned wisdom. Highly recommended!"
Douglas Brinkley
Katherine Tsanoff Brown Chair in Humanities and
Professor, Rice University
Author, *Cronkite*

"Ignition is a must-read for anyone who wants to harness the power of communication and understand the strategic process to make words more powerful and persuasive." Moseley not only shares with you how he achieved successes in the communications field but even more importantly the wisdom he learned from the obstacles he experienced. Ignition is the essential field manual for communications strategy!"
Rosalind Wiseman
Co-founder, Cultures of Dignity
NYT bestselling author, *Queen Bees &*
Wannabes (the movie Mean Girls)

"Moseley serves up an insightful, witty, and fast-paced guide to practical tactics and techniques for communicating better to the world around us. Told through a memorable cast of characters, his hard-earned wisdom will speak not only to communications professionals but also to organizational leaders and everyday people looking to make an impact with their words. Take a seat, grab a bowl of gumbo, and dig in!"
Colleen Scanlan Lyons, Ph.D.
Anthropologist, author, professor
University of Colorado, Boulder

"Ignition is a fascinating journey with Matthew Moseley as our guide helping us understand the relationship among words, actions, and music and how it all relates to us in ways that we don't just understand, but also FEEL. Like Marshall McLuhan long ago, Moseley opens up doors and takes us to new vistas where we can study how we are influenced by our surroundings and how we, in turn, can influence others. Using his considerable gifts as a storyteller, he has created an enjoyable book that makes you THINK!!! In addition to being a fun read, Ignition serves as a textbook for studying this world of communication and as a guide to making a better one."

David Amram
Composer, Author, *Pull My Daisy*

"Moseley provides little nuggets of wisdom about how we can better communicate with each other and ultimately be more successful in all endeavors, both personal and professional. Required reading for anyone managing an organization."

Suzanne Stoller
Director of People Operations, Google

"Matt Moseley is a true communications savant. In his newest book, Moseley pulls back the curtain on all things related to communications strategy — from the philosophical paradigms on which we make decisions to the granular details of execution. Ignition offers a masterful blend of story, cultural and historical examination, as well as the research behind how we communicate as individuals, companies, and societies. This book is a must-have tool for any organizational or community leader who is dedicated to truly influencing their audience."

Michael Diettrich-Chastain
Author, *Changes: The Busy Professional's Guide to Reducing Stress, Accomplishing Goals and Mastering Adaptability*
Founder and CEO, Arc Integrated

Ignition

Ignition
Superior Communication Strategies for Creating Stronger Connections

Matthew L. Moseley

In this age, in this country, public sentiment is everything. With public sentiment, nothing can fail; without it nothing can succeed. Consequently he who molds public sentiment, goes deeper than he who enacts statutes or pronounces decisions. He makes statutes and decisions possible or impossible to be executed.

Abraham Lincoln

Routledge
Taylor & Francis Group

A PRODUCTIVITY PRESS BOOK

First published 2021
by Routledge
600 Broken Sound Parkway #300, Boca Raton FL, 33487

and by Routledge
2 Park Square, Milton Park, Abingdon, Oxon, OX14 4RN

Routledge is an imprint of the Taylor & Francis Group, an informa business

© 2021 Matthew L. Moseley

The right of Matthew L. Moseley to be identified as author of this work has been asserted by him in accordance with sections 77 and 78 of the Copyright, Designs and Patents Act 1988.

All rights reserved. No part of this book may be reprinted or reproduced or utilised in any form or by any electronic, mechanical, or other means, now known or hereafter invented, including photocopying and recording, or in any information storage or retrieval system, without permission in writing from the publishers.

Trademark Notice: Product or corporate names may be trademarks or registered trademarks, and are used only for identification and explanation without intent to infringe.

ISBN: 9780367747817 (hbk)
ISBN: 9780367559427 (pbk)
ISBN: 9781003159513 (ebk)

Typeset in Minion Pro
by codeMantra

For our children,

Charles and Amelia

Contents

PART 1   Communication from the Beginning:
History and Philosophy

PART 3   Why

List of Illustrations

Acknowledgments

The concept for this book began as a talk in 2015 in Black Rock City called "Gonzo Communications." For the talk I had created a pamphlet called *Ignition* about getting off the sidelines and putting oneself into the story. Thus began the journey of this book from fire and dust.

No Hero's Journey would be complete without the appearance of the helper. The mentor. For me, Ryan Carrasco appeared as the Book Doctor. As the next several years rolled by and the manuscript developed, he brought an uncanny ability to tease out concepts and fit pieces together in just the right places. Over our two years of work hammering out sentences, paragraphs, and chapters, he became much more than an editor. He became a true *compañero* and friend. I am eternally grateful for his belief in The Dream.

As with any long-distance swim or other big endeavor, the support team is critical. Most are in the Cast of Characters in the opening. As the process increased in intensity and momentum, there were many people on the *Ignition Book Team*. My sister, Mary Moseley Lobdell, has always been there throughout my life and with my books it has been no different. She is a shrewd editor and treasured contributor. Author Michael Dietrich-Chastain saw the book in its infancy and kept providing encouragement and feedback. My gratitude to Colleen Scanlan Lyons who provided valuable insight and comments.

Rosalind Wiseman was a counselor and mentor through the book-writing process and always ready with incisive edits and an encouraging word. Kolby Ward, the Maui Wild Card, and longtime member of my support crew, provided key insights at just the right time. My colleague, Isabelle Deibel, a bright young graduate from the University of Chicago, was indispensable and I extend my deepest thanks to her for keeping me organized.

My gratitude to my editor at Routledge Publishing, Michael Sinocchi, and Samantha Dalton, the editorial assistant, who believed in this book (even during a pandemic) and saw it through to fruition. A big thanks to Dean Birkenkamp, also at Routledge Publishing, who gave me early guidance on the proposal and early stages of the book. Emma Juniper brought extensive professional wisdom, inspiring curiosity, and razor-sharp commentary. Curtis Robinson has always been there in the trenches and encouraged me to place faith in my own voice. The legendary composer David Amram somehow touches every project of mine. *Ignition* was no different. No one on the group virtual call will forget his live reading of the final two chapters.

My gratitude to Eleanor Williams, who expertly crafted the illustrations. Dave Farmer captured the essence of the *Ignition* concept in designing the logo. Nicholas Guthrie and Zachary "Cal" Hoffman also provided indispensable help along the way.

There were a handful of people at the original talk in the hot Nevada desert: Counselor and friend from Tel Aviv, Gad Reich; my cousin, Glynde Mangum; Tom Giovagnoli "Tommy G," who has always been a spark; Allen and Jennifer Blow from New Orleans.

Michael (Nooch) and Rhonda Antonucci were invaluable friends throughout the entire journey. Many thanks to our legal team, Tom Ward and Dru Nielsen. Jonathan Bartsch has been a longtime friend by my side as a trusty sounding board for ideas. My heartfelt gratitude for wise and observant comments from Gordon Riggle (University of Colorado Leeds School of Business and the President's Leadership Class).

Longtime neighbor Carol Byerly, Ph.D., is a historian specializing in pandemics and the 1918 influenza outbreak. I appreciate her sitting down with me numerous times to discuss the state of affairs during COVID. Mark Williams has found himself frequently by my side in the middle of the night in a kayak. He kept me grounded and focused through the writing.

Lastly, my eternal love and gratitude to my wife, Kristin Howse Moseley. While I might be a river, she is the rock. And quite the wordsmith as well. Thank you for the enduring love and the support for all of my crazy ideas. A handful of which become butterflies.

Matthew L. Moseley
Boulder, Colorado
October 1, 2020

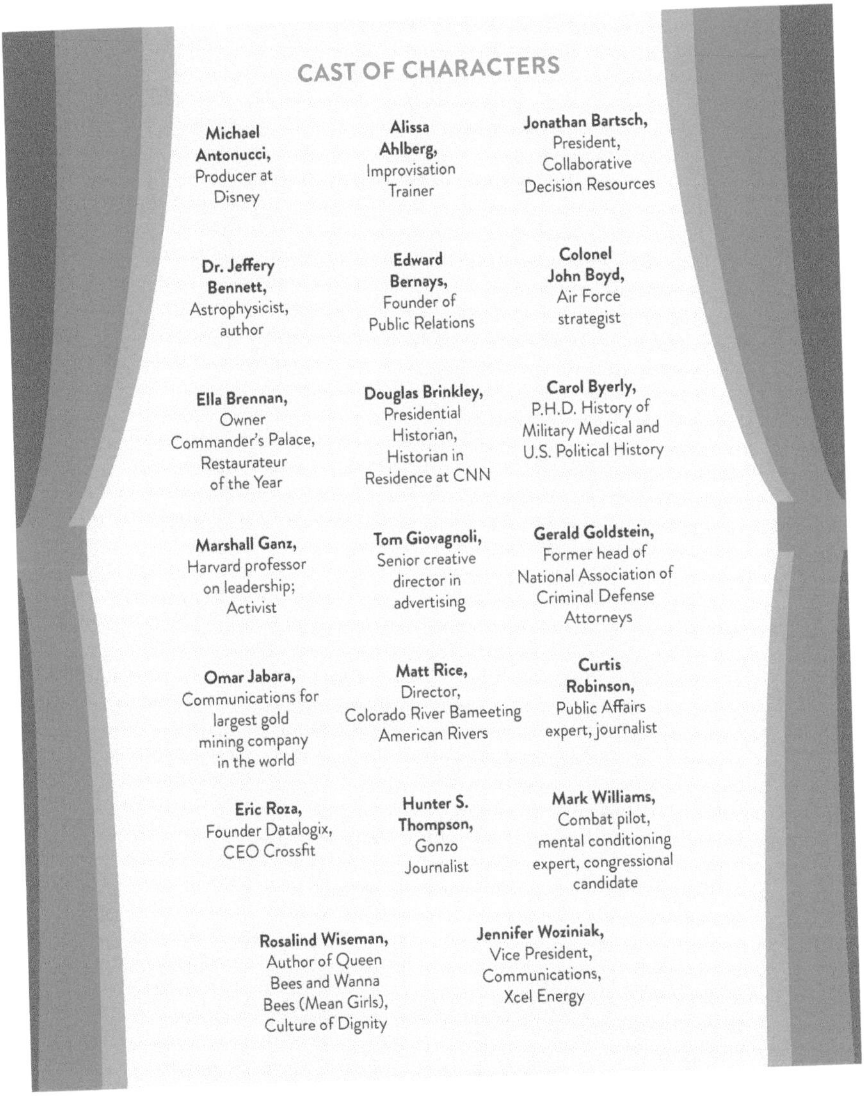

CAST OF CHARACTERS

Michael Antonucci,
Producer at
Disney

Alissa Ahlberg,
Improvisation
Trainer

Jonathan Bartsch,
President,
Collaborative
Decision Resources

Dr. Jeffery Bennett,
Astrophysicist,
author

Edward Bernays,
Founder of
Public Relations

Colonel John Boyd,
Air Force
strategist

Ella Brennan,
Owner
Commander's Palace,
Restaurateur
of the Year

Douglas Brinkley,
Presidential
Historian,
Historian in
Residence at CNN

Carol Byerly,
P.H.D. History of
Military Medical and
U.S. Political History

Marshall Ganz,
Harvard professor
on leadership;
Activist

Tom Giovagnoli,
Senior creative
director in
advertising

Gerald Goldstein,
Former head of
National Association of
Criminal Defense
Attorneys

Omar Jabara,
Communications for
largest gold
mining company
in the world

Matt Rice,
Director,
Colorado River Bameeting
American Rivers

Curtis Robinson,
Public Affairs
expert, journalist

Eric Roza,
Founder Datalogix,
CEO Crossfit

Hunter S. Thompson,
Gonzo
Journalist

Mark Williams,
Combat pilot,
mental conditioning
expert, congressional
candidate

Rosalind Wiseman,
Author of Queen
Bees and Wanna
Bees (Mean Girls),
Culture of Dignity

Jennifer Woziniak,
Vice President,
Communications,
Xcel Energy

FIGURE I.1
Cast of characters.

About the Author

Matthew L. Moseley is a communications strategist with decades of experience at the intersection of public policy, business, and government. He has managed many public affairs projects and campaigns for organizations and companies. He is the author of *Dear Dr. Thompson: Felony Murder, Hunter S. Thompson and the Last Gonzo Campaign*, which chronicles his joint campaign with the late journalist to free Lisl Auman from prison for felony murder. He has completed four first-ever record adventure swims and is the subject of the documentary, *Dancing in the Water*. He lives in Boulder, Colorado, with his wife, Kristin, and their children, Charles and Amelia.

Introduction

BLAST OFF

We are taking a journey together, not unlike astronauts strapped in and ready to launch. As we are about to take off, we recall those clarion moments of our past that have brought us to RIGHT NOW. I remember where my own journey began. I was standing on a stage in front of a full auditorium in Chicago in 1983, sweat falling from my brow. I was there because I had won the Louisiana State Catholic Debating Championships competition by outtalking everyone else in a state full of big talkers. But this was the big leagues now: a national stage with a sophisticated audience. I was a long way from the bayous of south Louisiana.

A gift for gab and a freakish obsession with current events of the day had taken me all the way to championship in Extemporaneous Speaking, but as I stood before a packed auditorium, words failed to materialize when I needed them most.

The event worked like this: you picked three topics out of a hat, chose one, and had a half hour to come up with a five-to-seven-minute speech, which you then presented to three judges. When it was my turn, I pulled three topics I knew absolutely nothing about. Ultimately, I went with *How is acid rain from Canada affecting America?* Or was it, *How is acid rain from America affecting Canada?* (Heck, I'm not even sure I knew back then.)

My introduction went smoothly, but as I moved on to Point #1, I was talking about pollution rising up to the clouds... and then my words slowly trailed off into a mumble. I went completely blank. Fueled by panic, my body became a furnace. I stared through my steamy glasses and words babbled forth without comprehension. *Acid rain is caused by... factories... in the country of... Canada... and, uh, it's... not good...* Groping, I leapt ahead to my conclusion.

I paced the stage and, with feigned conviction, said something like, *We must push the boulder of acid rain coming down from Canada over the edge. To finally topple acid rain and solve this problem once and for all.* I realize it must sound as ridiculous today as I'm sure it did to the judges then. I was relying — feebly — on the myth of Sisyphus, my go-to analogy for nearly every speech and debate competition. Sisyphus is the Greek God who was cast into a pit for his transgressions and forced to roll a huge boulder up the sloped sides. Every time he pushed it right up to the edge, the boulder would come rolling down upon him and he would have to try again, a never-ending task. I would relate whatever the topic was to Sisyphus' boulder and use it as a metaphor for that subject. His boulder could be neatly applied to any issue, from youth criminal justice reform to civil rights, from education policy to... well, acid rain.

In opening the speech with Sisyphus, I could come back around to The Rock at the end for the conclusion and tie up the talk in a nice bow. Usually, I could find a file of pertinent information on the public concern in my catalogue case, research the issue, and the speech would go over well. Not this time.

And this wouldn't be the last time in my life where words didn't materialize in a crucial moment. I've tasted acid rain more than once. I've witnessed the same embarrassment and shame in other people, close friends and colleagues, when they stumble. Where, instead of clarity, there is only confusion.

I lost the competition that day, but realized valuable lessons that have stayed with me ever since, and reaffirmed over a lifetime of communications experience. Namely, how to argue both sides of an issue, the power of metaphor and symbolism, and — rightly or wrongly — *the person who can tell the best story wins.*

I didn't know it then, but I had found my life's mission. My failure on the national high school debate stage wasn't the end I thought it was. It was just the beginning.

*****

From that debate stage, I went on to work on major campaigns, crisis situations, and ferocious public policy battles. Inside these pages we pull back the curtain to the secrets of working alongside CEOs, communication directors, presidential candidates, advertising geniuses, environmental warriors, campaign wizards, billionaires, small nonprofits, labor organizations, and a lot of lawyers. Most often, though, we work with normal everyday people just trying to make a difference in the world around them. Through these experiences, I've been scorched, baked, glazed, smothered, chunked, and deep-fried. I've been battle tested. And through it all, I've learned a lot about how people relate to each other. Throughout a life spent moving from one challenge to another, I've noticed that one thread has woven the whole experience together: *The unbridled power of good communication and storytelling.*

This book was hatched with one overarching question: *If communication is so important, and we are more interconnected than ever before, then why are we so bad at it?* My purpose in writing *Ignition* was to dispel all the ambiguity and confusion by providing methodologies for people and organizations to improve their approach to communications, create stronger connections, and ultimately be more successful.

In business, or in any relationship, communication is *everything* — with team members, investors, customers, vendors, and everyone in between. Ineffective communication results in lost moments and squandered opportunities. In our bodies when cells don't communicate properly it can indicate illness. When nations don't communicate with each other, it can cause war. Not communicating with your spouse can lead to divorce. Communication is the lubricant that keeps the engine of humanity running.

As a practical guide to more effective communication, this book consists of three parts. **Part 1** goes back to the beginning of humanity to show the intricate dance of art and science in all types of interpersonal interaction.

Part 2 takes us inside the brain of a communications strategist to unpack three vital strategic planning questions:

- *What* are you saying? (Messaging)
- *Who* are you saying it to? (Targeting audiences)
- *How* are you saying it? (Tactics)

In addition to sharing tried-and-true lessons from the field, Part 2 serves as a complete guide to tactical communications planning and includes the *Ignition Communications Template.*

Part 3 offers lessons and strategies from the front lines of communications. This knowledge culminates in a unique framework and vision for how we approach our world and each other.

Anyone who engages in communications — whether in business, organizational management, crisis communications, advertising, education, law, politics, science, or campaigns of every stripe — will find value in these timeless, universal truths. *Ignition* is about the art of controlling your environment through communications and making stronger, more meaningful connections.

The countdown is on. The time is Now.

4...3...2...

Ignition.

Also, by the Author

Dear Dr. Thompson:
Felony Murder, Hunter S. Thompson and the Last Gonzo Campaign

Part 1

Communication
from the Beginning:
History and Philosophy

1

Spark

Strange children should smile at each other and say, 'Let's play.'

F. Scott Fitzgerald

To understand the place of communication in human society, let's start at the beginning. You could say that the whole story of humanity is wrapped up in the evolution of how we communicate. From wordless grunts to hieroglyphics, the first phonetic alphabets to today's eBooks, humans have developed remarkable inventions for a single purpose: the transmission of ideas. In the quest to more perfectly share an idea or a thought with others, our methods of communication change, and with it, so too does society.

When we interact with others, we are attempting to establish a visceral connection. When we engage with someone intellectually, emotionally, or physically, we create a spark, a bond — something more precious than gold. When we communicate, whether it's speaking or writing, we do our best in choosing among all possible words, expressions, and gestures to pick those that most accurately match our intent, the ones that capture the exact message we're trying to send with special care given to how that message will be received. Unfortunately, perfect matches rarely happen.

In particular, quantum physics, which essentially studies the relationship between energy and particles, may have a special relevance to the mechanics of our social world and how we relate to each other. Whether it's having a conversation with your mother, waging a hard-fought political campaign, or placing a to-go order, our interactions do not take place in a vacuum — they have significance that ripples through time and space. Somewhere in the esoteric equations of quantum physics, there may be an explanation for the unforeseen, yet inevitable consequences of our communications.

Think of all the millions of words in all the languages of the world. Now think of the infinite combinations and permutations among them. Consider the countless ways that utterances can be modified by tone, facial expression, pitch, and volume. It's on the scale of the infinite combinations of notes and expressions of various arrangements of musical instruments.

Let us think, then, of each component of expression as a unit. Similar to a molecule or a living cell. The behavior of these units of expression, when combined, can be pat and predictable, perhaps like the valedictorian speech at a high school graduation. But these expressions can also be strange, mysterious, and even mystical.

When we combine words, gestures, and other signals, it all becomes greater than the sum of its parts. From here we can formulate a notion of *Quantum Communications*: the energy behind an act of communication is more important than the words, facial expressions, and gestures that comprise it. In physics, a "quantum leap" refers to the sudden transition a particle makes from one level of energy to another. It is abrupt, with no in-between state. Here or there. One or zero. This is what we're after in our communications: the energy that ignites us to *do something*. To believe. To care. To go from one state of being to another.

We have inherited communication from distant evolutionary ancestors, so we engage in it instinctively, not always understanding the logic. Communication is like particles of energy that we exchange. As we slide and swivel past each other, this exchange functions as currency for attraction and success.

In 1939, Dr. Adolf Butenandt was awarded a Nobel Prize for discovering and isolating human sex hormones. Later he pioneered investigations into what attracts insects to each other. It wasn't until 1959 that Dr. Butenandt finally discovered the secret ingredient of silkworm romance: the pheromone bombykol. In his extensive experiments, he harvested just 6.4 mg of bombykol, less than a thousandth of an ounce, from more than half a million silkworms. Even this minute amount was enough to arouse half the males of any sample.

Bombykol communicates the desire for sex; thus, it helps to perpetuate the silkworm species. But bombykol is just one pheromone in one species; how do pheromones work in general from silkworms to humans? Pheromones emanate from one animal and register with another, but neither may understand why they are experiencing attraction. Desire creates intention, which may inspire action and cause a reaction, but how these intentions and actions and reactions all come together exactly is a mysterious alchemy.

Humans, like almost every other living organism, rely on pheromones for behavioral cues. We involuntarily give off and receive these complex chemical compounds through scent, taste, and other receptors. Sometimes referred to as *ecto-hormones* for the way they trigger reactions in others, pheromones constitute an entire communication system, a catalyst for species perpetuation. As fundamental as pheromones are in serving our most basic instinct — reproduction — communication, extrapolated to its broader applications, is just as elemental to any notion of success.

CARING

Let's take a look at human behavior from a different angle. On a basic level, all mental and physical activity can be attributed to synaptic transmission. Synapses essentially send messages that tell your body to do something. To feel. To move. To create. To care. The energy passing through synapses elicits an all-or-nothing response by neurons in the brain and is a massive undertaking. Scientists estimate we possess around 100 billion neurons, each in contact with approximately 10,000 others. An impulse reaches these neurons, either from within the brain or from elsewhere in the body, and it either triggers a response or fails to do so. With speed that puts even today's most advanced supercomputers to shame, a stimulus sets off a chain of micro-reactions that, when taken all together, offer us the basis of all decision-making: first impressions, hunches, and intuition.

Science is beginning to unravel these complex processes in the brain, which distill a potent mix — our personal story, our familial history, the culture we were raised in, the culture we currently operate in, our emotions, physical/geographical location, and so much else — into a single conclusion or concept. We come to a fork in the road: either invest in something or let it go. Our own personal quantum leap — here or there? Ultimately, there is one criterion that drives all our choices: *Do we care or not?* This question informs how we interact with others and profoundly impacts how we make decisions.

Despite its binary nature at the granular level, the decision-making process becomes much more complex in the aggregate. Caring is not a science, but an art — which is another way of saying it's tricky. Care too little and you relinquish the invaluable ability to tell your own story, to advocate for yourself and others. Ennui often results. Care too much, however, and you strangle your environment, which leads to micromanaging everyone and everything. This can lead to anxiety (for you and everyone in your orbit), distraction, loss of productivity and creativity, stress, and strained relationships.

There is caring in the sense of becoming personally invested in something, and then there is caring in the sense of looking out for others. Naturally, these two senses often overlap. In their book *The Art of Happiness: A Handbook for Living*, the Dalai Lama and Dr. Howard C. Cutler offer wisdom that straddles religion and science. Cutler, a psychiatrist, adds expertise in the science of human happiness to his co-writer's spiritual wisdom. They reflect,

> Reaching out to others may be as fundamental to our nature as communication. One could draw an analogy with the development of language which, like the capacity for compassion and altruism, is one of the magnificent features of the human race.

The message to readers is to turn their focus outward instead of inward, to be more selfless and less selfish — in both word and deed. This turning outward is pertinent to our purposes because we're discussing communication. It is not enough that we are clear about what we wish to convey. We must also look out for our audience to ensure that our message is just as clear to them.

Scientists have now identified areas of the brain that are devoted to the *potential* for language, specifically Broca's and Wernicke's areas, which play a large role in production and processing. Even though humans may be hardwired for language and communication, this ability doesn't develop automatically. Language must be learned. The Dalai Lama claims that the human potential for compassion is similar — he even helped found a research center at Stanford University that investigates the development of compassion. He believes that, given the right circumstances and conditions, the seed of compassion will take root. With practice, we can develop the part of our brain responsible for compassion and caring — much in the same way our brain develops language proficiency. Fail to practice compassion and that part of the brain will stagnate. Here again, communication and compassion are directly linked since compassion gets lost if we are not able to express it to one other. Compassion ultimately depends on communication.

Cultivated by understanding others, compassion is a fundamental component of effective leadership. Recall the old adage about putting yourself in someone else's shoes: if you can connect with and understand people, you have an unquantifiable advantage. As we will demonstrate, this advantage is critical to being successful in a crisis situation. To build a house, raise children, feed the homeless, or run a Fortune 500 company, we must have compassion and empathy.

Of course, we should also consider whether compassion is relevant for someone in a position of managing a budget or the harsh realities of a business closing down. While a person may feel compassion, this emotion alone will not prevent layoffs and factories being shuttered. Let's shift the focus from others to ourselves for a moment. "The purpose of life is happiness," says the Dalai Lama unequivocally. "The turning-toward happiness as a valid goal and the conscious decision to seek happiness in a systematic manner can profoundly change the rest of our lives." We must ask ourselves, Do I want to be happy or not? If you do, you must do something about it: humans must express needs and desires. To be happy, we must communicate.

If compassion and happiness call for communication, communication calls for self-awareness. In Plato's Phaedrus, Socrates compares the tripartite soul to a charioteer and two winged steeds. The charioteer is our reason. One horse represents our rational, moral side, and the other represents our irrational, impulsive side. I simplify the analogy to illustrate our consciousness (or awareness). This version of the analogy grants us more agency by putting the reins directly in our hands. Let's say we're the chariot driver and as such, it's our duty (and our great challenge) to keep our two horses — one of reason, one of emotion — moving in unison. Too much emotion or too much reason and one of the horses runs off the road and our chariot winds up in a ditch.

So, what does all this have to do with communication? Everything. Communication is about the control of emotions and the management of reasoning. The very act of communicating begins with regulating emotional reactions and responses, matching them with what we care about the most (our most essential and meaningful priorities), balancing logic and desire, and finally determining how best to use language to share what we have distilled. Research shows us that self-management of emotions is the most important skill in determining achievement and success. This control, our chariot driver, is what regulates our communication with the outside world as well — a concept emphasized in the stories throughout this book.

People can usually sense when another person is happy or angry. Those who are particularly aware or know us well can sense our mood right away, even from a distance. My wife, for instance, can read me the moment she sees me and knows exactly how I'm feeling. Even if I say nothing, she knows when something is bothering me, because mostly I'm fairly jovial. Some may describe it as intuition. The field of organizational development calls it *emotional intelligence.* My friend, Michael Diettrich Chastain, who is a therapist and the author of *Changes: The Busy Professional's Guide to Reducing Stress, Accomplishing Goals and Mastering Adaptability*, believes people can build this capacity to interpret others' emotions and train for it in much the same way one trains for a marathon.

There is good reason to train. Happiness is a contagion. It begets more happiness. Regrettably, the same can be said for anger. If you show people anger, they'll respond with anger. Bad vibes breed more bad vibes. It's a law of human nature, appearing in the proverbs of Ancient Egypt, underpinning such maxims as "The Golden Rule," and drawing points of comparison among the teachings of Jesus, the Buddha, Confucius, Muhammad, Lao Tzu, and even the Existentialist philosopher, Jean-Paul Sartre. Social psychologists call it the "Law of Reciprocity." Neuroscience has even identified so-called "mirror neurons" in the brain, which account for certain responses, such as our tendency to smile back at someone who smiles at us.

ENERGY

If words are the warp and weft of communication, the energy behind them is the loom. Unfortunately, sometimes in spite of ourselves we send the listener in the wrong direction. We must continually *disambiguate* — or clarify — our communication. We often do so instantaneously. Considering the countless ways to convey a message — words, tone, and volume, for example — how can we be sure that what we have chosen is the most unadulterated expression of our intention? That we say what we really mean to say? That our words match our intention? That our words match our body language? The form of communication matters just as much. When we want a message to stick, circumstances will determine if having a face-to-face conversation will be more effective than placing a call or sending a note. Texts have the potential to be misinterpreted because they lack tone, pitch, and other qualities that capture the full extent of a message's complexity. To mitigate this misunderstanding, emoji designers have tapped into the nuance of emotion to convey sentiments that many would find difficult to express in words.

Verbal communication and written communication have different advantages and limitations. In one respect, verbal communication has the advantage of being enhanced by inflection — and, if we can see the person — facial expression, and gesture. Then again, spoken language can be ephemeral. As soon as speech stops ringing in our ears, it becomes a ghost, subject to distortion in our error-prone memory or lost to oblivion altogether. (Written communication, on the other hand, is more permanent by its nature as it can be referred back to repeatedly.) Though permanence is more difficult to achieve with spoken content, we tend to remember what someone said to us that triggered a particularly intense emotion. As dynamic speakers know, if they neglect to engage audience members emotionally, they soon will lose them.

Despite the different characteristics of the various forms, every communication is a form of energy released into the ether. Just as tiny pulses provide the necessary energy for all cellular function, so too does communication supply power to all human activity. Oratory, advertisements, songs, literature, and art transcend time and space and stir emotions in new audiences. Think of the great speeches throughout American history. Dr. King: "I have a dream..." President Kennedy: "Ask not what your country can do for you..." President Lincoln: "Four Score and seven years ago..." Or the great books that have influenced and inspired humanity. They encapsulate so much more than words. Even more than mere ideas. They illustrate the effect that the principles of quantum physics have on communication. A *quantum communication*. They live on through new generations because they continue to connect hearts and minds. These combinations of words influence lives and culture. They represent leaps in human progress and shared understanding. They are not just products of historical context; they make history, they create the future.

RELATIVITY

I have spent much of my career advising clients on communications strategies. I am not a physicist. So, to better understand my hypothesis that a combination of words and signals can become greater — or lesser — than the sum of its parts, I turned to fellow swimmer Dr. Jeff Bennett, who also happens to be a world-renowned astrophysicist. I admire Professor Bennett for his deep understanding of the universe and for his dedication to help others deepen their understanding as well. He is a pioneer in space education. His book *Max Goes to the Moon* was the first book ever to be read in space as a part of NASA's "Story Time from Space" program. In 2014, Dr. Bennett published a groundbreaking book revisiting Einstein's Theory of Relativity called *What is Relativity?: An Intuitive Introduction to Einstein's Theories, and Why They Matter.*

Our conversation made me think about applying the Theory of Relativity to the act of communication. Einstein's theory posits that our perspective depends upon where we stand at a particular point in time and space. In a nutshell, Einstein concluded that the laws of physics are the same for all observers. He discovered that space and time are interdependent, and that there is a continuum known as space-time. The implications of this discovery continue to astound us. Theoretically, at least, events that occur at one time for one observer could occur at different times for others.

In terms of communication, this corresponds to the idea that the same message will mean different things at different times and places. As the context of a message changes, so too does the message for each sender and recipient. Theoretical physicist Richard Feynman said that context is everything — even in physics. Nothing can exist without its context. A message might mean one thing the first time we see it and something completely different the fifth time. We will continue to explore this relationship (one that advertisers and marketers spend a lot of time thinking about) in Part 2.

My discussions with Professor Bennett also led me to investigate the antagonistic force of *Entropy.* The Second Law of Thermodynamics states that the universe has a general tendency toward disorder. The very essence of a biological organism, however, is to create order. Imagine a character out of an old Western trying to create a good life on the rugged and hostile plains or Elsa in *Frozen* seeking clarity about who she is and her powers. Such ambition in the face of adversity doesn't happen by accident. Neither in the movies nor in real life. Order requires intention, and it also requires energy. To create order, any organism or system must communicate, which happens through the exchange of energy. *But where does this energy come from? Our heart? The sun? The universe? A god?*

These questions got me thinking about schools of fish and the many other species of animals that move together in a self-organizing group. In the research paper, "The Self-Organizing Quantum Universe," published in *Scientific American*, authors Jan Ambjørn, Jerzy Jurkiewicz, and Renate Loll compare the flight of European starlings to a dynamic at work in quantum physics. Starlings simultaneously moving together in a form of small-scale organized chaos (a phenomenon called *murmuration*) nevertheless demonstrate order and trends. Each bird is only in proximity to a few other birds. No leader tells them what to do. Yet they move together as a whole. The many become one.

Human behavior and cultural movements operate similarly. We move according to a *zeitgeist*, a German term that means the "spirit of the times." Formed by our values and norms, the zeitgeist is essentially the collective mood that we come to associate with a particular historical era. The point being, we are self-organizing creatures. True, we have leaders, but while they may tell us what to do every now and then, ultimately we make our own decisions. The same principles underlying flocking behavior, schools of fish, and even the causal nature of the universe are also present in people. We are all starlings self-organizing into some greater whole. (I've learned that the concept of murmuration can also be neatly applied to the flocking behavior of teenagers.)

Most of us think of a system as having fixed dimensions. A profit-and-loss report or a cost/benefit analysis assumes a classical space, that which is used to describe a physical system, such as a box or a road. But after talking to Dr. Bennett I began to consider another view. I studied systems that function in a *Hilbert Space*, a space whose dimensions are limitless. The points within it, moving all around, are not fixed in time or space. A quantum particle, as opposed to a classical particle, doesn't have a precise position or precise momentum. This is explained by the Heisenberg Uncertainty Principle, which, simply put, states that the more certainty with which we know the location of an object, the less certain we can be of its velocity, and vice-versa. Essentially this means that classical concepts, such as position and momentum, can only be used in approximate ways when applied to a quantum system. They are impossible to predict with complete confidence because the laws of quantum dynamics are often "random" or probabilistic, not deterministic.

Unpredictability is intrinsic to the world itself. Consider weather, earthquakes, the stock market, elections, and the whims of dictators. This is precisely what makes quantum theory so apt in describing communications. There are no deterministic laws. Quantum mechanics may as well be describing a political campaign or an elementary school classroom as particles. We are all creatures moving around with different moods and impulses at any given moment. The way we exchange signals varies based upon our context, our history, and the way we anticipate an audience receiving our message. There is no certainty.

But... and this may be the most important information I share with you: if we develop certain skills, we can zero in with greater accuracy on matching intent, articulation, and audience. We can never know exactly what someone *will* do, but we can determine what someone is *likely* to do. (Think public opinion polling.) This sharper focus becomes paramount when we're engaged in a crisis situation or running an advertising campaign or managing a big public policy issue.

We can develop the muscles of communication just as the Dalai Lama insists we can develop our compassion and caring. Our mission is not just to create something greater than the sum of its parts, but to develop the strength to communicate what we've done to the world. If we adopt a quantum perspective, anything is possible.

2

Controlling Your Environment

Five years before his death, I worked closely with the late Gonzo journalist Hunter S. Thompson. Our time together was full of unforgettable memories, but one moment stands out above the rest. Late one night, as we sat in his kitchen outside of Aspen, Colorado, he took a drag on a Dunhill cigarette in his signature filter and out of the blue asked me, "Do you know what the definition of the word, politics, is?"

I had worked in politics and government for many years, but that night I stumbled on this most basic question. Overwhelmed by the Herculean labor of cramming so much into one quaint phrase, I finally just rattled off a list of buzzwords: *voting, government, campaigns, civics, civil society,* etc.

Nope. Not according to Dr. Thompson. "All wrong." He drew on his cigarette and exhaled slowly. With utter authority he said, "Politics is the art of controlling your environment."

Hunter explained that the candidates and organizations who thrive and win are those who tell better stories and control the message. (What makes the best story? Good question. We'll get to that in Part 2.) They draw the media to *their* events and press conferences. It meant framing the tone of the debate and exchange. He explained that *Controlling Your Environment* means being proactive and setting the agenda, doing things to make others respond to you, rather than vice-versa. People and organizations who are on the offensive are defining themselves rather than having others define them or, perhaps worst of all, settling for no definition whatsoever.

This is as true today as it was when Hunter said it (and whenever the same basic point was made before him!). Especially in the fast-lane of today's digital era, those who control the narrative and tell a better story carry the day.

That night Hunter looked at me, took a sip of Chivas, and emphasized the philosophy behind *Controlling Your Environment*, "When you're explaining, you're losing." If you have to explain yourself or your organization and answer questions about what you should have done, if you have to respond to an accusation, then you're already on the defensive. If, Hunter conjectured, Richard Nixon had alleged that his possible Democratic opponent, President Lyndon B. Johnson, had an affair with a young woman: maybe it happened, maybe it didn't, but either way, Johnson would have to respond.

It's important to take control whenever we can. Consider for a second all the elements under our control. We can turn the thermostat up or down. We can train to run marathons and dream up long-distance swims. At least, in most free and democratic societies, we can choose our partner, tell our boss to "*Take this job and shove it!*," and publish our ideas. In general, we get to choose what we put in our bodies. We can pick a margarita or a glass of wine. We can decide what kind of music to listen to, what we read, where we go, and with whom. Even children, who enjoy far less freedom than do adults, still have plenty of leeway. As observed by Anne Frank, who managed to do something extraordinary while trapped by circumstances beyond her control, "Parents can only give good advice or put them on the right paths, but the final forming of a person's character lies in their own hands."

Hunter Thompson wasn't the first to articulate the concept of being proactive. In business, there is a term called *First Mover Advantage*, which refers to the upper hand a company gains in launching its products first, dominating a particular segment of the market, and establishing brand recognition and customer loyalty before competition arises. Generally, the First Mover in communications has more opportunities to set the terms of the debate, define the terminology and language, and frame the issue. Yes, there are risks — sometimes massive, career-ending risks — by getting out there first. The First Mover is opened up to scrutiny, criticism, and a position for others to attack. But those attacks are usually *on your terms* and *using your language*. Those attacks are reactions. Consider, as an example Tesla, the leader of the electric vehicle industry, setting parameters that traditional carmakers in the old internal combustion engine model (Ford, GM, Fiat Chrysler) must now take into account if they're to catch up.

While the concept is most commonly used in the context of technology products, it reinforces Hunter's maxim: *You need to make others react to you.*

Of course, there are also things that are beyond our control. We don't get to choose our parents or our schoolteachers. We can't control the paths of hurricanes. Most of us have little control over terrorism or gang violence or others' aggression, for the most part. No one prefers to be a refugee. I can't imagine anyone choosing to be born into a fascist dictatorship. The point is while being proactive beats being reactive, it's also better to prepare for those instances in which circumstances — some crisis, for example — force you to react, rather than to delude yourself into thinking that you're directing matters when you're not.

What is essential for any good communicator is an acute sense of how much control is possible to assume at any point in any situation. In general, we have a higher degree of control in times of certainty and economic success, and a lesser degree of control in times of distress, poverty, and pandemics (which is why opportunists and provocateurs usually rush in to capitalize on fear.)

In regard to controlling our environment, awareness is the most important tool. We simply can't shape our world if we don't know what is going on or don't have a clear idea of what we want our world to look like in the first place. Awareness is our starting point for how we shape our world. (In the next section, I'll take you through different methods and perspectives on developing your awareness so that you can make better-informed decisions.)

In this light, perhaps it's useful to revisit the phrase, '*Controlling Your Environment*.' Does it come up a bit short in accounting for the all-too-turbulent nature of our reality? Perhaps it would be better expressed as *trying* to control your environment. As any sailor knows, the craft is at the mercy of the sea, and we do the best we can to navigate the currents with the information we have at present. Perhaps an even more accurate expression would be *Shaping Our Environment*, molding events and reactions to those events so as to suit our objectives. Even as Hunter conceded, politics, communications — life itself! — it's all an *Art*, not an exact *Science*.

This leads us to perhaps one of the most important concepts in this book: the *Locus of Control*, a psychological concept regarding how one approaches different situations in life. It is also closely related to how one creates meaning and interprets events. One's perspective is largely determined by whether one's locus of control is *internal* or *external*. People who have an external locus of control say things like "I'll get to go to Carnegie Hall if it's *God's will*" or "My *destiny* will guide the way." They believe they are bound by forces beyond their control. My mother was very much this way. But if a person says, "I'm going to Carnegie Hall, and *I'm going to make it happen*, no matter what," he or she is exhibiting an internal Locus of Control. As they say, the only way to get to Carnegie Hall is to *practice, practice, practice.* Those who fail to finish marathons and blame it on their coach or the extreme heat exhibit an external locus of control. But if you blame yourself because you partied half the night before or didn't do the proper distance training or simply came up short — that's an example of an *internal* Locus of Control. We are all capable of both internal and external viewpoints, sometimes simultaneously. What is important is realizing what your *Locus of Control* is when making choices. With all due respect for the wide diversity of psychological profiles out there, we are most successful when we know we have the right to tell our story and control what that story is. We can choose how much it opens us up. Or closes us down.

Some say most of life is beyond our control. When Albert Camus wrote *The Plague*, he explored many themes in addition to disease, but part of what the book meant was that life is perishable. For any reason, at any moment, our lives can be extinguished by an infinite amount of causes. The ultimate lesson that Camus was conveying, is that this makes every day that much more precious and meaningful. Controlling what we can and managing the rest.

Linguistic theorists, such as George Lakoff, and a long list of philosophers before him, would assert that our perspective on our environment is the most important factor in how we make sense of the world. If we *think* we will fail, then we are more likely to give up and fail. If we *believe* we are one of the smart kids, then we often become the smart kid by seeking resources and spending time to make it so.

Dr. Amy Cuddy is a social psychologist at Harvard who delivered one of the most successful TedTalks of all time, "Your body language may shape who you are," which has over 57 million views. Her big idea is that "Power Posing" — standing in a posture of confidence, even when we don't feel confident — can boost feelings of confidence and have an impact on our chances of success. Her research has revealed that we can change how people perceive us, and perhaps even our own body chemistry — simply by changing our body position. The way we communicate nonverbally reveals much more about us than do our actual words. Think about how people make sweeping inferences and judgments from *nonverbal* communications that have profound consequences. Who we hire? Who we hang out with as friends? Who we collaborate with on our most important projects? We are constantly put in situations where we are judging others, others are judging us, and — most importantly — we are judging ourselves.

This opinion of ourselves, which determines our Locus of Control, is reflected in our body language — *nonverbals*. These cues are easily recognizable to the trained eye. Dr. Cuddy naturally gravitates toward asking questions about power and dominance. She explains that people who are confident, who have won a big race or feel enthusiastic or victorious, appear big. Winners raise their hands high in the air. People who are feeling powerful often stretch out their arms. They take up more space. They might cross their legs in a meeting and drape their arm over a nearby chair. This body language shows pride and confidence.

Dr. Cuddy gave the example of students in her classrooms. The smart, confident ones sit near the front, have their materials spread out, and raise their hands high and with authority when answering a question. They tend to speak louder and without a lot of *hmmm*s, *uhh*s, and *like*s.

It's just the opposite with those who feel powerless and afraid. When people lack control or awareness, they make themselves small; they contract with shoulders slouched. They might rub their neck or play with their hair. This is what happens when people are glued to their cell phones and constantly checking social media. Think about when you're feeling depressed or lonely. You want to curl up in a ball.

Dr. Cuddy wanted to know if you could exhibit body language that demonstrates power even when you're feeling powerless or like a fraud and what happens when you do. And, if that were possible, would those body signals then influence how you think of yourself in power situations. Her question was simply, "Can you fake it till you make it?"

Her conclusion was *Yes*. In power situations it is often not the competence of the person that is the ultimate determination of success, but the impression that the person gives to others.

My sister, Dr. Mary Lobdell, Ph.D., is an expert speech pathologist with decades of practice with patients. She makes use of the principle, *Fake it till you make, it* all the time with her patients and clients. For young children learning to communicate, a primary motivation is trying to establish control of their environment. My sister will deliberately put treats like M&Ms in a child-proof container so that her young patients will have to resort to language and ask for help. She will stand in a child's way so that he or she will have to ask her to move. The children are forced to learn how to communicate to exert control. She applies the same principle with elderly stroke patients whose verbal skills have been compromised. She starts them off with basic words like "help" and single words for foods they enjoy, such as "rice," and then prompts them to use the words to get what they want. Both of these therapies help patients control their environment by compelling them to communicate their needs to their caretakers.

Our *Loci of Control* reflects how we view our place in the world and is on constant display through nonverbal communication. Or, as Dr. Cuddy illustrates, we can make a conscious choice to display something different. Her simple, yet profound, finding is that *The body can shape the mind.* I would contend that her theory is an affirmation of a universal truth. The ability of the body to affect the mind might be at the very root of human survival and dominion.

POWER

The Godfather, Part III, may not have been the best of the franchise, but it had a potent tagline: *Real power can't be given. It must be taken.* It was referring to power over turf, over people and transactions. Political activist Gloria Steinem, a far cry from Michael Corleone, expressed something very similar, "Power can be taken, but not given. The process of the taking is empowerment in itself."

No discussion of communications would be complete without addressing the slippery notion of Power. Ideas have power. Voice is power. A platform is power. Some think power is getting elected to office or making a lot of money. Political positions and account balances are temporary states of power that can and will change; true power lasts. So, let's talk about *maintaining* power. Holding on to power requires using positions and resources to acquire a deep comprehension of what other people need and expect — and then coming up with an appealing story that helps assimilate that knowledge into action.

There are two basic kinds of power in the world: power that is bestowed on you through the privilege of position, wealth, status, and social connections. And then there is power you bestow on yourself. Agency. Gumption. Enterprise. Ambition. The power of realizing a vision by expressing it to others. The power of assembling a coalition. The power of inspiring people through music, images, and rhetoric and mobilizing a political force. The power to create something of consequence, something greater than the sum of its parts.

Having worked with many CEOs, elected leaders, and directors of organizations, I've discovered that the best have a heightened sense of the interests of those around them and can relate to those interests by communicating a vision. Power comes from understanding — and then managing — interests, which can vary widely even within a small group. Without such an understanding, a communications vacuum is created.

Leadership and power are close cousins. Marshall Ganz, whom we will return to often, is a leadership professor and theorist at Harvard. He wrote an influential piece for the Harvard Business Review called *Leading Change*. Ganz explains, "Leadership is accepting responsibility to create conditions that enable others to achieve shared purpose in the face of uncertainty." He continues, "Leaders accept responsibility not only for their individual part of the work, but also for the collective whole. Agency, however, is more about grasping at possibility than conforming to probability." Money alone cannot substitute. (Just ask Michael Bloomberg about his $300 million presidential campaign that went nowhere.) A keen understanding of interests and a superior ability to manage relationships make up the secret sauce of good communication and leadership.

In his book, *Never Split the Difference*, FBI hostage negotiator Chris Voss, writes, "Politics aside, empathy is not about being nice or agreeing with the other side. It's about understanding them." The better you understand someone's emotions, the more power you have when interacting with them. Voss has particularly good advice to never think of someone on the other side of the bargaining table as an "enemy." On the contrary, to create more power, develop a relationship. Ultimately, this approach produces the best outcomes. Sufficiently understanding interests requires a deep sense of empathy, a true curiosity about how someone else approaches the world. Seeking to understand not just *what* someone across the table is saying, but *why*. Empathy is an easy concept to understand theoretically, but it is much harder to *practice* in the real world. This is difficult for many people precisely because they are usually so concerned with what they themselves are going to say. They are focused on how they are being impacted. This is ego, not empathy. True empathy helps us to better understand the audience we need for a product, a production, a ballot measure, or campaign for elected office.

These muscles of empathy and understanding also power us through our everyday lives with confidence and clarity. Empathy reduces the chance of misunderstanding, diminishes anger, and assuages the fear that accompanies confusion. So, if only for your own sake, strengthen your empathy and compassion. Clarity and understanding will occur more naturally. Don't expect an *Aha!* Moment. It's more of a slow development of quiet confidence, a keener understanding of how to approach the world. You will be more likely to accept the power and responsibility of an internal locus of control. To generate empathy for someone or something, ask these three questions:

What are their interests, values, and norms?
What do they expect from us?
What is the reaction to that expectation?

These ideas of controlling our environment, empathy, and shaping perspectives are the most important seeds of wisdom in your garden.

3

Meaning Out of the Mire

Let's consider the physiology of the voice box, our earliest and most primary communications tool. Most people think that the purpose of the larynx is for talking — it is not. The biological function of the larynx is to protect our airway, which is why it is situated at the top of the windpipe. When you swallow properly, the larynx closes to protect the windpipe. When something goes down the wrong way, you cough. Swallow wrong and you can die. This is how the voice box developed.

The vocal cords are surprisingly small — only about a centimeter and half long — particularly small for something that can make so much noise. The larynx was formed as a survival mechanism, not a communications tool. One theory contends that when humans were just beginning to stand on two feet, communication was limited to grunts and growls. People could only express basic emotions like love and approval or threaten violence and with little subtlety. They mostly expressed what they felt with little emotional control, similar to what we see today on certain social media accounts.

Language evolved and written language was developed along with the domestication of plants and animals, when people accrued excess agricultural goods to trade with others. Our ancient ancestors needed a more sophisticated way to discuss and record their transactions. They needed to be able to communicate specific information about their merchandise, such as where it came from, its value, and terms of exchange. They needed to measure and quantify their transactions. Language was the way to express value.

The recording of human thought was a game changer for humanity. With it, we could go back, analyze our thinking, and revise it. We weren't just animals shrugging and grunting anymore. We could express greater complexity than could be conveyed via smoke signal or even pictographs on canyon walls. We could begin to share the best practices for the common good.

Yuval Harari, author of the sweeping history of humankind, *Sapiens*, explains how this capacity to think outside ourselves, to create myths and art, is our X factor. This ability paved the way for humans to organize into large groups and connect by way of a common purpose and meaning. Once language created the ability for humans to think in conceptual ways and unite around these concepts, we could communicate the sense we made of the world and thus organize the tribe and the kingdom. Harari introduces us to the oldest piece of figurative art ever discovered, a figurine called the "Lion-Man" or "Lioness-Woman" that was created about 32,000 years ago and found in the Stadel Cave in Germany in 1939. The head is leonine with the body of a human. Harari asserts, "This is one of the first indisputable examples of art, and probably of religion, and of the ability of the human mind to imagine things that do not really exist."

In addition to explaining how long this impulse to broadcast our ideas has been with us, he also illustrates just how effective it can be in uniting us as a people. He cites the *U.S. Constitution* as a document in which Americans organize an entire political system of common thought and purpose. Before that was the *Magna Carta*. These principles, and their interpretations, have become the playbook for an entire nation. But, despite their historic significance, this artifact and this document are still only two ways of framing the world.

Communication is how we relate to our world to build common connections and manage the environment around us.

SEAT OF MEANING

The media wasn't always so disparaged. Journalists were not always vilified and scorned. In fact, it used to be just the opposite. Early scribes were revered as mystics. Holy diviners. They gave order and clarity to the exchange of goods and services, and by doing so, facilitated the transmission of knowledge. They created history by chronicling events and made sure their greater significance was passed down through the generations.

They conferred meaning through stories, some of which we still read today. In certain circles they were shamans, illuminating holy wisdom and presenting visions. At one time, writing was a closely-guarded secret. Tales and legends were hidden on scrolls, imprinted on canyon walls, and etched on stone tablets. Power came from the ability to interpret, distill, and clarify in a confounding world.

When these wise ones communicated, they were sharing. The word, communicate, comes from the Latin verb, *communicare*, which is defined as to make common a thought or to share an idea. It is also closely related to the Latin word for "common": *communis*. Think of "common sense" in the most general of terms. When we communicate to the public, we contribute to shared knowledge and understanding. Communications scholar Karl Erik Rosengren calls this contribution "the basic precondition for all community." If we don't communicate, we cannot have commonality. We would be lost souls wandering through a barren landscape — no tribe, no shared resources, no protection — trying to survive on our own.

In earlier times, oral communication was the key to power. Without persuasion, there was a vacuum. There was no movement. During the Greek and Roman empires, in centers of learning like Rome and Alexandria, scholars developed the *Trivium*: grammar, dialectic, and rhetoric.

This elegant little triumvirate forms the ancient art of discourse. Together, these three Hellenic columns represent an effort to resolve the ambiguity of speech and cultivate greater understanding. *Grammar* includes the foundational rules that dictate the ways words become phrases and nouns and verbs leap into sentences, and sentences are the sites on which meaning is constructed.

Dialectic is the process by which parties with differing viewpoints come to an accepted truth through various methods of engagement and debate.

Lastly, there's *rhetoric*, the timeless art of getting people to care. Rhetoric is the means by which we inform, persuade, and inspire others.

These three columns of the *Trivium* meet at a confluence that defines how we express our values and beliefs in a river of expression. Our needs and desires rage through rapids of reason. Our thoughts of love, friendship, and our place in the world swirl in eddies of emotion. The *Trivium* is a channel for the everflowing stream of the mind. A Ship of Meaning floating above the deep sea of confusion.

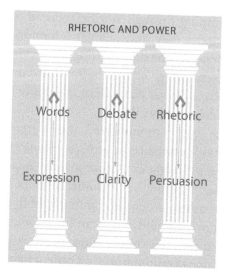

FIGURE 3.1
"Rhetoric and Power."

Even today, people read the speeches of the Roman philosopher and politician Cicero to improve their own command of language. In Ancient Athens, "sophists" purported to teach anyone how to win any argument. For a price, that is. To slander a philosopher was to accuse him of sophistry, essentially calling him a charlatan and a swindler. But such were the times. It calls to mind dusty Grecian imagery: early philosphers sitting in togas all day under stately Doric columns, pondering oratory, speech, and man's relations to the heavens and each other.

Scholars generally agree that language developed with the Cro-Magnon Man period around 30,000–50,000 years ago. It's hard to know exactly when oral language developed, however; there wouldn't be any record left behind. But we know that around 3500–3000 BCE written language appeared in southern Mesopotamia by the Sumerian. Along with the creation of written symbols came symbols representing numbers. In the 8th century along came Greek alphabetic numerals, but it is believed this system has Egyptian origins dating back to the 6th century. Basic math, calculus, algebra, and statistics soon followed. Scholars trace back to 6th century India the watershed moment when zero first took its place as a digit. It would take a while to realize the full potential of this new language. The notion of zero and one, as in the binary system of synaptic transmission — on and off — has become the basic structure of computer science, which operates according to a universal language all its own: "code." It's a language beyond semantics. Code is the New Latin of our times. It has spawned an entire cultural phenomenon. Silicon Valley. Google. Artificial Intelligence. Apple. Social Media and all the attendant ways of life that have changed with it. All of it is based on communications.

Remember the first single-celled organisms? When the first printing press was invented, it caused a similar revolution. Humans, like those first organisms, opened eyes to a whole new world of interconnectivity. We could ask questions to others about the nature of ourselves. We could look into the heavens and seek answers about *our* place in the universe. Understanding led to more understanding. In the 17th century, newspapers began to flourish and a seismic shift happened in the way information was shared. The proliferation of the printing press not only presented scholars with an unprecedented capability to share their works, it also eventually allowed just about everyone to get their ideas out to the larger world. It wouldn't be long before Steve Jobs was dropping acid and dreaming of a revolution in how we relate to each other. Wouldn't you know — it was a communications device under the guise of a simple apple with a bite taken out of it.

As we discussed in Chapter 1, communication doesn't happen in a vacuum. At its simplest, a group comprises two people — called a dyad. (Three people make up a triad, four a quartet, and so forth.) The difference between each may seem small, but it is much more significant than the incrementalism may suggest. A triad is 50% larger than a dyad. When the number of communicators (*n*) in a system grows, the number of potential direct relations within the units of the system expands. Thus, communications scholars have developed a fairly simple formula for looking at group communications:

$$R = \frac{n(n-1)}{2}$$

You may recall from math class that this is the formula for calculating how many different pairs you can draw from a group. Consider a group that has five members: $(5 \times 4)/2 = 10$ possibilities for two-way connections between the five members. Double the group and there are more than four times as many possibilities: 45, to be exact. And with the increasing complexity of group structure (*quantitative change*), the communicative system of the group undergoes *qualitative change.* What starts as communication between individuals in a group mushrooms into communication between and within every combination of groups within a system.

Here you have the science behind an idea *going viral.*

But this is just a formula for analyzing communications. As communications have a quantum structure, we can predict, but we cannot preordain. Even with our best estimate, it is like catching lightning in a bottle. Very smart people spend millions of dollars on advertising campaigns guaranteed to succeed by experts, focus groups, and preliminary trials... only to watch those can't-fail campaigns fizzle upon launching. The introductions of New Coke and Bud Dry come to mind. And then there's the kid who sits in his basement, makes a YouTube video about making a pizza with gummy bears or slime, gets millions of hits, and kicks off a lucrative career based on instant internet celebrity.

Any person who looks a client in the eye and makes predictions about events with absolute conviction is masking inherent uncertainty. Sometimes language doesn't register with an intended audience as expected. Oftentimes there are intervening events. Competitors unveil a secret weapon. There's an employee crisis. A strike. Weather can derail the best-laid plans. A global pandemic shuts down the world for an indeterminate amount of time. Carl Von Clausewitz, the famed father of military strategy, who battled Napoleon, called it *friction* (more on this in later chapters). The best leaders have good plans, but their real skill might lie in how they manage this friction moment to moment in the ongoing drama of everyday maneuvers.

Just how do we go about understanding the world around us? We experience *phenomena* and use *concepts* to understand them. All that we experience on a daily basis — traffic, phone calls, hikes, bus rides — are phenomena. Every moment you are not asleep you are experiencing phenomena. These experiences and interactions register in our consciousness as *concepts*, internal story bites that help us make sense of our world. Part of developing concepts is to make distinctions.

When we feel, we react — and through our reaction, we give *meaning* to a phenomenon. This can only happen by creating a link in the mind. Charles E. Osgood discusses this very thing in his 1957 groundbreaking book, *The Measurement of Meaning.* Specifically, he considers the dimensions of meaning and applies a statistical technique he calls *factor analysis.* Osgood concludes that we define our reality and the phenomena around us by a surprisingly small number of dimensions. These include:

Evaluation (good/bad), Strength (strong/weak), and Activity (active/passive). These, and only these, three criteria form the legs of a stool that holds the seat of meaning. While we may imagine all these complex emotions and thoughts swirling around the gumbo pot of our mind, when we boil it down, there are only a few simple ingredients. We *evaluate* a situation, first and foremost, by asking ourselves, *Do I care? How much? And what, if anything, will I do about it?*

Communication cannot happen without meaning. Communications scholar Karl Rosengren suggests that "Above all, communication concerns the process of meaning creation: questions concerning how people create meaning psychologically, socially, and culturally; how messages are understood intellectually; and how ambiguity arises and is resolved." Meaning is co-created; it is a social construct. For the communications strategist, there can be no meaning without interpreting actions, events, and reactions. Our crucial task then is to understand this process by which meaning is created.

Viktor Frankl wrote about this very thing in his book, *Man's Search for Meaning*, which details his experience as a prisoner in Nazi concentration camps. Frankl writes about the tremendous power — for all of humanity and for the individual — of defining our own events and creating our own meaning rather than being a passenger to our existence by having our own meaning defined for us. Frankl said we are the framers of our universe. Us, not some other person, government, ally, or adversary.

FIGURE 3.2
"The Seat of Meaning."

Our ability to make meaning out of just about anything can be put to various ends. I'm always amused to consider how much meaning can be squeezed from the humble grape. There is a global community of vinophiles based upon the creative use of adjectives to describe how a little grape can become such a complex work of art. The wine industry is very inventive about words to describe certain flavors. Words that evoke experiences. You will hear *notes of vanilla, lemon curd, and ripe apple*, but you will never hear *it tastes like a very good grape*, which, at its essence, is what it is. An entire industry is based on adjectives…and well-grown grapes. Is it Brazilian dark chocolate with tamarind and a grapefruit finish — or is the wine really just a well-fermented red grape with a nice label, catchy name, and the right price?

The wine industry gives us a good excuse to break down *words*. All text is "unreliable." As words take, at least, some of their meaning from their relationship to other words, their full meaning is always in flux (a fact that lends language its *quantumness*). The context is what confers meaning — there is no correct or objective way to understand a given word. After dinner at our family cabin in the deep woods of Bayou Chicot, Louisiana, with cicadas singing and Spanish Moss draping the cypress trees, Mary explained to me how linguistic theory can broaden our understanding of communications.

She gave an example: You have a ball. It's round and made of rubber. A tangible object. However, the *word* for the ball is a different thing. It is not tangible. The word, "ball," is not the actual ball — only language we use to signify "ball." The word is a construct of the mind to conceptualize the ball, which is useful when talking about a ball with others, who themselves have internalized the concept of *ball*. Harari contends that our ability to share a simple concept gave rise to our ability to organize for common purposes. To connect the actual object and the word for it, your brain creates an *index*. The index is critical because it directs your mind from word to concept. An index is your mental map. This may seem simple, but when nailing down the goals and the direction of an organization or a campaign, creating an index can be shockingly difficult.

Mary suggests thinking of the term, *index*, as an extension of your index finger literally pointing in a particular direction. She said when people physically point to an object, like a river or a ball, they are emphasizing and strengthening the index.

The question is always: *Is an index being used correctly?* An index is generally used to clarify. But it can also be used to obfuscate. Consider legal jargon or the inscrutable instructions on the long form of a tax return. Think of the Terms and Conditions for consenting to use a social media platform like Facebook or Twitter. The "fine print" seems to exclude all but the lawyers who write it or the accountants trained in deciphering IRS code. Think of the obscure language used by academics. Those who aren't well acquainted with this language lack a reliable index to their meaning.

Applying linguistic theory to politics, we can determine when politicians obscure an index in their speech and when they create clear direction. Candidates for elected office send out position statements designed to appeal to the broadest readership and to avoid offending anyone (I know because I've written them). There is a lot of euphemism and reading between the lines. For instance, when working with labor organizations we don't use the term, "union," we say, "working families." The term, "union," can be laden with preconceived notions, such as corruption, union busting, mafia ties, and strikes, but who can argue with families just working to make a living?

Frank Luntz, the legendary Republican pollster, and Karl Rove, senior advisor to President George W. Bush, were masters of this strategy. Think about phrases like "Clean Skies" initiative" (which would have reduced air quality standards) or "No Child Left Behind" (the overemphasis on standardized tests left many children behind). How about "Mission Accomplished" (when the Iraq war was just beginning)?

I have a lot of respect for Luntz, who is a master of using language — even if his use relies on subtle shifts in indexicality that can be misleading. I respect him not because of his content or his ideas, but for his techniques. (He would probably praise effective left-leaning messaging for the same reason.) He is the author of *Words Matter: It's Not What You Say, It's What People Hear.* Luntz has famously suggested Republicans use the phrase, "climate change," instead of "global warming." He argued that instead of referring to an "Estate Tax," which is literally a tax on an estate, use "Death Tax." Luntz describes his specialty as "testing language and finding words that will help his clients sell their product or turn public opinion on an issue or candidate." He is a wizard at taking complex policy issues and boiling them down to soundbites and catch phrases. This is much harder than it appears on the surface.

Then there is one of my heroes, James Carville, who is also from South Louisiana, who coined one of the most inventive and elegant phrases of all modern political campaigns, *"It's the economy, stupid."* A very clear and decisive index to the meaning and thrust of Bill Clinton's first campaign and the zeitgeist of uncertainty from an economic recession.

My old friend, Jared Boigon, a top political consultant in San Francisco, once told me at the Madrone Art Bar on Divisadero Street, "Words can be the *enemy* of meaning." I asked him what he meant by this. He said we may feel joy, elation, frustration, love, and anger — even all at once. *Confuzzlement.* Where do we find the vocabulary to express all the complexity of what we feel at a particular moment, much less the totality of the human condition? Sometimes people don't know what they want to say. The heart is sending our brain signals, but we don't always know how to interpret these emotions or express them even when we do. Emotions are rarely unalloyed. They are often a mix. Like *bittersweet*. Or love and jealousy in a relationship. This is why poets, screenwriters, musicians, and artists of every stripe should command such respect. It is difficult to capture the potency of our compounding and complex emotions. When it happens, it can move our souls and change our lives. Good language should inspire and create common purpose and meaning out of the mire. A shared index.

DISPLAY RULES

I have a hypothesis: *Of all the elements of human nature that contribute to our happiness and survival as a species, communication is the most fundamental.* This is because our beliefs and values are anchored in *phenomena, concepts,* and the *meaning* we create, which is our *index.* Now we come full circle back to happiness. If we applied what we now know about these linguistic theories and the Seat of Meaning to the Dalai Lama's declaration that the meaning of life is happiness, it would sound something like this: We can only be happy with a *phenomenon* after evaluating it, determining how much it was good or bad, and deciding to care or not. Only then can you decide by how much.

People unknowingly operate according to what communications theorists call *Display Rules:* culturally prescribed norms that govern how people express themselves and socialize. For instance, in some places it is polite to burp after a meal or slurp your soup, and in others, it is not. These cultural norms set the general boundaries for a society regarding when and where to express emotions and to what extent.

How cultures greet one another is a great way to think about display rules. *What's up, Bro?* A curtsy for the Queen. Bowing to Your Excellency. Saluting the General. A kiss on the cheek. In general, display rules apply to interactions, both verbal and nonverbal. And their impact is vast. They influence everything from who we interact with to what we wear. They also have a major impact on policy and government — think of the decorum that we usually expect our elected leaders to abide by. There are also display rules on ball fields and in classrooms. Display rules are rigid in a board of directors meeting. But they are also rigidly enforced at popular surf breaks, where they dictate who gets what wave and when. If you don't understand the norms and values of a tribe, then you can't possibly gain entrée and, in turn, anticipate, much less influence, how members of that tribe will perceive an issue or a product. This is extremely important in creating advertisements that reinforce and play off of cultural values and relevance. In other words, understanding display rules is critical to success in communications.

Think for a moment of all the people you come into contact with every day. It could be family, friends, or members of your hockey league. Maybe you're socializing after church, seeing someone about a permit for an addition to the house, or dealing with the government for a new driver's license. The interaction could be good or bad. Helpful or frustrating. Maybe all at the same time.

There are remarkably few institutions that actually create the rules and norms for how our society operates. Among them I'd include schools, workplaces, places of worship, and sports. Such institutions are in the continuous process of communicating their practices and values. Grandparents, parents, and peer groups pass along the traditions of communities. New meaning is created as the old groups and practices give way to new generations and innovations.

Our consciousness is a 24-hour bakery, kneading the ingredients of our world into the bread of meaning.

4

Public Opinion: A Brief Primer on Strategy & Public Affairs

I've worked with dozens of clients over the years: Global telecom giants, environmental organizations, labor unions, energy companies, education associations, alleged criminals, and everyone in between. One of the joys of my professional life is helping people tell their stories. Defining a perspective. Crystallizing what it all means. My work has sent me on some epic adventures. From expanding the franchise through music at Rock the Vote to staging presidential events within the White House Communications operation to delivering first-class hospitality at America's top restaurant. Partnering with environmentalists, CEOs, celebrities, politicians, heiresses, and even a few dirtbags — they, too, have their own unique story. I've worked with the biggest and the smallest.

I've had a lot of time to think about strategy. What worked and what didn't.

Strategy... Oh, what a word. Lofty. High-minded. Mysterious. Conceiving of the world as a chessboard manipulated by a few bigwigs in smoky back rooms. Wizards behind the curtain. Strategy is a word so overwrought that it has been rendered pretentious. *Don't you worry, leave the strategy to us.*

That's one way of looking at it. Of course, despite its overuse, strategy is still vitally important. Just the act of thinking about something, breaking it down, and making a plan is a thing of beauty in and of itself. *Strategy* is The Plan, the 360° view. It's how all the pieces fit together.

Ok, but what *is* strategy? Let's pull back the curtain on these wizards.

The simple word, "Strategy," can be added to anything to make it sound more planned and thought-out: strategic thinking, a strategic communications plan (redundant), strategic cycling (a spin class), strategic relationship building (cocktail party circuit). Marshall Ganz says "Strategy is how we turn what we have into what we need to get what we want — a hypothesis that if we do *x*, *y*, and *z*, then *a* will result."

A lot of people have a lot to say about strategy. Sir Lawrence Freedman is one of the world's preeminent international scholars on politics and war. His 2013 book, *Strategy: A History,* which took 40 years to write, is the best book ever written outlining the entire history and development of *strategy.* *The Economist* called the book "Magisterial." Freedman writes, "Having a strategy suggests an ability to look up from the short term and the trivial to view the long term and the essential, it reflects a capacity to see woods rather than trees." Humans naturally long for this kind of clarity when chaos and confusion reign. Freedman says,

> So the realm of strategy is one of bargaining and persuasion as well as threats and pressure, psychological as well as physical effects, and words as well as deeds. This is why strategy is the central political art. It is about getting more out of a situation than the starting balance of power would suggest. *It is the art of creating power* (emphasis added).

So, when a client asks, *"What are we doing here, folks?"*, we are the ones who'd better have an answer, and preferably a well-crafted document.

Saul Alinsky, hero of the labor movement, a patron saint of union organizers, and author of *Rules for Radicals,* wondered how to use these same concepts of high strategy to wrest power from the oligarchs and tycoons and give it to the people. Like many revolutionaries before him, he wanted to share the wealth with the workers. Alinsky's contribution was twofold. First, he adopted the tools of those in power and used them against them. Second, he maintained a laser-focus on strategy and tactics. Alinsky was a master at community organizing. He believed that real change would come from the bottom up. He fought against corporatism, gentrification, and the creation of new slums. Among his many famous actions, in 1961 he bussed 2,500 black residents of Chicago to the polls to vote.

There are different interpretations of strategy. Beverly Gage is a history professor at Yale University, where she directs the Brady-Johnson Program in Grand Strategy. In a September 3, 2018 *New York Times Magazine,* Gage goes back to the beginning: "The original concept of strategy comes from the world of military affairs. It derives from a Greek word meaning 'generalship' or 'the office or command of a general': it was an enterprise for the man in charge."

One of the most recognized books on strategy is from the British military historian B.H. Liddell Hart. In his book, *Strategy*, he defined the concept as "the art of distributing and applying military means to fulfill the ends of policy," distinguishing strategy from, on one side, "tactics" — the modes of "actual fighting" on the battlefield — and on the other, "grand strategy," in which civilian leaders set high-level policy and coordinate the nation's resources toward a collective goal.

There is a temptation to confuse a vision or policy with a strategy, but they are not the same thing. Policies address the "what." They're prescriptions for the way things might operate in an ideal world. Strategy is about the "how." How do you move toward a desired end, despite limited means and huge obstacles? A policy may have an implementation strategy behind it. Strategy is often associated with high-level decision makers — generals, presidents, corporate titans — but the basic challenge of, in Theodore Roosevelt's words, doing "what you can, with what you have, where you are" applies just as much when working from the bottom up. Strategy is agnostic about who is currently in power. Alinsky believed the right strategy would allow anyone to wield power.

Any strategic challenge requires contending with limits and obstacles: scarce resources, structural constraints, devoted enemies and fickle allies, chance, and luck. The Plan is a thoughtful prediction.

WHAT THE PEOPLE SAY

Communication Strategy encompasses many aspects and fields, public relations being but one subset. There are the cousins of advertising and marketing (where the focus is mostly on selling a product or service to a consumer); and then there is branding, reputation management, issue management, government affairs, ballot measures or referenda and citizen-led initiatives, political campaign management, and public affairs (my speciality). Even government relations (a nice word for lobbying) is targeted communications within a subset of elected officials or regulatory agencies.

Some form of public relations has been at play ever since people started gathering in tribes. People have always wanted to market (and mark) themselves. Especially in regard to attracting the best mate (we're not content to let pheromones do *all* the work). There has always existed the Great Game of acquiring some kind of respect and authority. Therefore, there likely have always been those who contemplate the power of reputation and optics and advise leaders accordingly. These are folks who give a lot of thought to words, symbols, music, and art and how to use them to influence people. Be it university deans, shamans, scribes, court jesters, chiefs of staff, or the old frail guy behind the massive curtain. Lord Varys, the Master of Whisperers in *Game of Thrones* reminds us, "Power resides where men believe it resides. It's a trick. A shadow on the wall. And a small man can cast a very large shadow."

But just how did the modern public relations profession begin?

Tracing public relations back through the 20th century brings us to the term, *propaganda*, which, having been deployed to devastating effect by leaders of all stripes in times of war and peace, has taken on sinister and pejorative overtones. The term propaganda wasn't always so vilified. The word had its beginnings in the Catholic Church reaching out to those who had not yet converted. It was also part of the Counter-Reformation, the Church's effort to oppose the first wave of upstart Protestantism. The standard definition is "for the propagation of a particular doctrine or practice." The term became much more disturbing after the Nazis brought it to terrifying new levels of effectiveness. During wartime, propaganda essentially consists of deliberate lying to raise troop morale, comfort the citizenry, and confuse the enemy.

As the United States entered the First World War in 1917, President Woodrow Wilson established the Committee on Public Information (CPI) to control messaging and shape public opinion on the danger of German militarism and the need for a response. Before long, the CPI would expand into other realms of public policy and suppress information and criticism of the government during the 1918 influenza epidemic.

Wilson installed George Creel as head of the CPI. A hard-driving, progressive journalist from Denver who cut his teeth at the *Rocky Mountain News*, Creel famously observed, "people do not live by bread alone: they live mostly by catch phrases." Creel hired some of the best and brightest from around the country — academics, advertisers, journalists — and brought them to Washington D.C. for what would become the first official propaganda machine in the United States.

The CPI was organized into five divisions. The Speaking Division recruited the "Four-Minute Man" to give brief speeches at civic organizations, like the Rotary Club. Seventy-five thousand people across America received talking points and bulletins about why people should support the war.

The Film Division made short films for theaters to be shown before main attractions as part of a film program. These short films almost invariably championed the superiority of American soldiers and the heroism of the Allies in their fight against the barbarous Germans.

The Foreign Language Newspaper Division monitored and influenced the hundreds of English newspapers around the world.

The War Bond Division was charged with promoting and financing the war effort by encouraging Americans to buy bonds. They hyped investment in U.S. bonds as an act of patriotism for the greater good — not just for U.S. interests, but for the freedom of the entire world against the forces of evil.

The Division of Pictorial Publicity produced iconic imagery to motivate citizens to support the war by appealing to a sense of civic duty and pride. This division produced the legendary "Uncle Sam" poster, "I want you."

FIGURE 4.1
"I Want YOU."

There are various perspectives on the utility and morality of propaganda. Critic of the CPI, journalist Walter Lippmann, who was influenced by John Dewey at the University of Chicago, closely followed the psycho-analytic movement's study of how people develop consciousness and why they act rationally or irrationally. Lippman coined the term, "stereotype," and the phrase, "Cold War." In 1922, Lippman published his landmark book, *Public Opinion*, wherein he argued that what people understood and what they knew about the world around them was just a "picture in their heads" of a "pseudo-environment." The problem, as Lippman explained, pertains to the stakes involved. He said, "But because it *is* behavior, the consequences, if they are acts, operate not in the pseudo-environment where the behavior is stimulated, but in the real environment where action eventuates."

Or, as the Chicago sociologist William Thomas put it a few years later, "If men define situations as real, they are real in their consequences." To a large extent, we decide what is real by bringing the pictures in our heads into the world. Lippman noted that people relied on a "system of stereotypes" because it provided a more "ordered, more or less consistent picture of the world, to which our habits, our tastes, or capacities, our comforts and our hopes have adjusted themselves." Because of this, he said, "any disturbance of the stereotypes seems like an attack upon the foundations of the universe." All of this meant that public opinion was suspect. Lippman believed, in practice, that public opinion was the construct of democratic consent and, therefore, could be manufactured.

Harold Lasswell, who would become known as the father of U.S. political science, made his mark with a theory of propaganda. He proposed that an understanding of the mobilization and manipulation of public opinion was paramount in foreign policy and particularly in armed conflict — domestic or otherwise. What once could "be done by violence and intimidation must now be done by argument and persuasion." Aristotletian rhetoric. As such, Lasswell acknowledged that the primary challenges for the strategist were "to intensify the attitudes favorable to his purpose, to reverse the attitudes hostile to it, and to attract the indifferent, or at the worst, to prevent them from assuming a hostile bent."

The CPI only lasted 18 months, but it would have a profound and lasting impact on how Americans viewed themselves. It reinforced a sense of national identity and collective purpose. Despite its best intentions, however, it also left behind a legacy of xenophobia and jingoism. The CPI was also active during the 1918 Influenza Pandemic suppressing information about the real harm of the virus. It forbade newspapers from criticizing the government's handling of the crisis. It threatened newspapers with a crime under the Sedition Act, which incidentally was used by Abraham Lincoln to shutter over 400 newspapers during the Civil War. The Sedition Act can still be used to narrow and direct public opinion today. For our purposes, the CPI may be most important because it was the womb from which the field of public relations was born.

Edward Bernays, nephew of Sigmund Freud, set up the first public relations shop in 1919 following his time at the CPI and was the first to use the term. Bernays was an optimist who believed that techniques for influencing public opinion could be used to better society. He considered a *Press Man* and a *Counsel on Public Relations* two distinct types that drive public opinion. A Press Man is a liaison to the media and is always trying to get free coverage. A PR Counsel creates news. Bernays set out to frame public relations as a highly respectable profession with a history rooted in social sciences and psychiatry. He contended that the interests of a complex society — from governments, corporations, non-profits, labor unions, and organizations all trying to gain an advantage — called for expert advice in how to communicate effectively.

His first book, *Crystallizing Public Opinion*, published in 1923, followed on the heels of his mentor Walter Lippmann's *Public Opinion*. In 1928 Bernays published his second book, *Propaganda*, with the opener, "The conscious and intelligent manipulation of the organized habits and opinions of the masses is an important element in democratic society." Those opinion-makers are "an invisible government which is the true ruling power of our country." As a result, "we are governed, our minds molded, our tastes formed, our ideas suggested, largely by men we have never heard of." Bernays knew what he was talking about. He was one of those men.

When Bernays died in 1995, he was described in his obituary as "the father of public relations." He was later named by *Life* magazine as one of the 100 most influential Americans of the 20th century. Bernays argued for a strict ethical code for the public relations profession and that the needs of society as a whole come before those of any individual. He advised clients on their complete relationship to their environment and distinguished himself from advertisers and marketers, whom he portrayed as pleading with people to purchase a certain commodity. Bernays recognized the "Dark forces from below" and argued that "the duty of the higher strata [is to] inject moral and spiritual motives into public opinion." He stated plainly, "Public opinion must become public conscience."

Just who is this master and "higher strata?" Ah yes, here is where it gets complicated. There are many competing opportunists and provocateurs — with their own agendas and visions — looking to exploit situations and opinions. It's a tricky question, but a very important one: the *who* and *what* of the so-called higher strata that try to guide the *zeitgeist*. Or, as Alinsky argued, is the "lower" strata just as powerful in this respect?

In an influential article, "The Engineering of Consent," Bernays applies the military metaphor, *tactics*, to the strategy of public relations. In writing about tactics, he encouraged the use of public spectacles to create news because these events could be communicated to "infinitely more people than those actually participating, and such events vividly dramatize ideas for those who do not witness the events." Bernays knew how to create a stunt. In 1929, he followed his own advice promoting the American Tobacco Company in New York City's Easter Parade, when he orchestrated ten debutantes on a float to light cigarettes simultaneously. By undermining the taboo against women smoking in public, he fed into the rising feminism of the age. The cigarettes became "Torches of Freedom."

Bernays pointed out propaganda was all around us. As our strategy scholar Lawrence Freedman notes, Bernays believed that "People and groups had a right to promote their ideas and the competition in doing so was healthy for both democracy and capitalism." The alternative is that people are not allowed to express ideas openly and there is no marketplace for the currency of communication, which inevitably looks like fascism and totalitarianism.

THINKING DIFFERENTLY

There is a supreme irony that in this time of profound communication technology advancement, most people feel more and more alone. Across the board, statistics reveal a heightened feeling of isolation. How can we be so *inter*connected, yet feel so *dis*connected? Why are the world's interpersonal networks breaking down, and what changes must we make to ensure that they meet our basic needs once more? It just might be that technology is rendering old communications rules of thumb increasingly irrelevant. In doing so, however, it also opens new realms of possibility.

Think of the invention that has done more to change the world in the last 20 years than almost anything else. The iPhone, a profound little communication device. The entire swath of Silicon Valley is built upon communication networks. Google, Facebook, Twitter, Instagram, and all the others are essentially communication companies. They are just the latest incarnation. The heyday of radio broadcasts emanating from New York, and then television newscasts constituted communication revolutions. Look at Hollywood. It's nothing but one giant storytelling factory. Artificial intelligence is on the rise, which is another way of saying artificial communication. How long until we humans become obsolete and our opinions are formed and expressed for us by technologies and algorithms we aren't even aware of?

One of Bernays's central questions has always intrigued me: *How are people persuaded to think differently? How can we get people to care about something — anything!* His questions inspire my own. How can public policy — so often dry, distant, and ambiguous — be presented so that it engages the very public it claims to serve? How can we get people to care about education funding, protecting rivers, healthcare, or stopping the deforestation of Amazon rainforests? One might even argue that the overall purpose of the public relations profession is to create relationships and make people *happy*.

So, if we are to form a causal link between action and happiness, we need to do more than just smile. This is where public affairs, my area of expertise, comes in. A specialized field within public relations, public affairs is more about the management of issues like water, education, and transportation. It may involve a whole array of related fields, such as digital advertising, branding, government relations, events, and coalition building. Public affairs normally occur at the intersection of public policy, business, and government. The *affairs* are the external issues facing an organization that have the potential to impact them in some form or fashion. It could be a ballot measure, a bill in the legislature, an angry mob of citizens, or inquiring media. Public affairs span the public and private sectors and can be complex. As we shall soon see, they may constitute threats to an organization — but just as often they present opportunities. Public affairs is the alchemy of turning money and resources into power.

5

Rule of Reaction

On Jeopardy! Alex Trebec once prompted, *In the Sapir-Whorf hypothesis, this ability of humans shapes how we view reality.* If you had blurted out "Double-jointedness!" or "Taco Tongue!" you would've lost and not just because you forgot to phrase your response in the form of a question. The correct answer: *What is Linguistic Relativity?*

The Sapir-Whorf Hypothesis of linguistic relativity is a principle that claims that our language structure can determine our worldview. In other words, our perspective and perceptions are relative to our spoken language, which sounds a little like what my astrophysicist friend Jeff Bennett said about applying the Theory of Relativity — the context of the sender and receiver are always changing in relation to each other. Our understanding of our environment might be a matter of, not just where we sit and when, but also how we speak.

There are two principal ways to view this hypothesis. The *Strong Version* holds that language is the primary determinant of thought and that our limits of linguistic categories are the boundaries of our cognitive abilities. On the flip side is the *Weak Version*, which stipulates that these linguistic categories only *influence* our thoughts and decisions. The principle has been debated since the early 20th century. Noam Chomsky said Linguistic Relativity was all rubbish. People could think whatever they could conceptualize and beyond. Language was just a vessel to express it, but certainly didn't negate it.

As I was writing on my front porch one summer afternoon below two ferns, I saw Dr. Al Kim, a good friend from the neighborhood, riding by on his bicycle. I hollered for him to stop and join me for a glass of wine. Dr. Al is an associate professor at the Institute of Cognitive Science at CU-Boulder in the Department of Psychology and Neuroscience. In his research, he investigates the cognitive and neural mechanisms that allow humans to understand language. He explains that communicating with human language is a signature capacity of our species. While humans understand words and sentences of their native language with relative ease, this belies complex neurological processes that allow this understanding to occur. Primarily, Dr. Al studies *verbs*, which constitute the foundation of action in grammar. "Verbs are complex because they express a thought within a thought. Not just the notion, but 'what are we going to do about that notion?'" Dr. Al says. "Verbs are the anchor of language." (Let's keep this in mind for later when we discuss creating messages and advertisements.)

"The process of language is extremely difficult," Dr. Al said. "The reason is that all words and expressions are riddled with grammatical ambiguity." I asked him to elaborate.

> It is much easier to be the speaker, to unleash a torrent of words upon a listener. This is why listening is much harder. We spend a lot more energy processing words and sentences as a listener than we do saying them as a speaker.

He told of how good listeners then translate this ability to knowledge to help them craft better stories and messages that, in turn, increases the capacity for someone to care about something.

Dr. Al explained that language is so complex because in order to accurately process the words of another person, we must first model that person, and what they are capable of understanding, in our own minds. This is the beginning of neural resonance and understanding what your listener expects. Neural resonance, essentially a vicarious imprinting on our own neural pathways, allows us to process what people are thinking and feeling. These reflections form the basis for empathy. "This is the pathway to comprehension, but it's not easy," Dr. Al said. "In fact, our research has shown just how complex this process really is." This development occurs in children about the age of four or five. Neural resonance is one of the hallmarks of human development. Instead of just "ball," children start to express more complex thoughts, such as "Daddy, let's go *throw* the ball." In this sentence, they model their minds on their father's and express a desire for a particular outcome.

We poured another glass of wine as the sun was starting to set and Dr. Al echoed my sister, Dr. Mary, by referencing the Theory of Mind, the ability to understand and attribute beliefs, emotions, intentions, and desires to others and to ourselves. Without such a theory, we wouldn't be able to understand others or determine that their values and perspectives are different than (or similar to) our own. By acknowledging the world-view of others, we come to realize our unique worldview.

When we aren't able to understand others or we have a deficit in com-prehending others' feelings and emotions, it can cause real physical and mental distress. It may be a symptom (or cause) of schizophrenia, atten-tion deficit hyperactivity disorder (ADHD), eating disorders, alcoholism, cocaine addiction, or a wide range of other illnesses and conditions.

Dr. Kim also referenced the *Sapir-Whorf hypothesis*. Like Noam Chomsky, he believes that whether our limitations of language are strong or weak may not matter because the very act of language itself is what drove human evolution. Dr. Al described how the human larynx is low in the throat. But he noted this came at a cost because the lower position makes us more susceptible to choking. When the larynx dropped, it gave us remarkable control and flexibility to create sounds and manage mes-sages. All of this is to say, starting around 150,000 years ago, we became hardwired for language.

Dr. Al posed an interesting question: Let's imagine you never learned a language. What would you do?

You'd make one up.

As it turns out, deaf children in Nicaragua were brought together in large numbers under a new public school. The children ranged in age from 4 to 16 and had no experience with sign language. At the school, they developed Nicaraguan Sign Language, a spontaneously created language.

I asked Dr. Al how someone could make use of his research on verbs to be better at communicating. He thought long and hard, staring down into the last few sips of wine. He finally replied. In his research into how the brain processes language, he has concluded that the real question isn't how to communicate *better*, but how we are able to communicate at all. Ambiguity is everywhere. Dr. Al said to look no further for it than in everyday newspaper headlines.

FIGURE 5.1
Ambiguous newspaper headlines.

TARGET PRACTICE

Shaping our environment isn't necessarily about being the best or the brightest; it's about handling the situation around us in whatever manner circumstances demand. If we focus too intently on being great and outstanding, we may become obsessed, blind to the bigger picture. We may be swept away by our emotions. We may chase shiny objects or tilt at windmills. Though our goals may start as a means to an end, an obsession becomes an impediment when it makes us lose sight of our goal. This happens in business and sports when people have a difficult time letting go of failing investments and end up chasing losses. *Maybe it will turn around? Oh God, please let it turn around…* Some people have trouble letting go of relationships. But let's be clear: Controlling your Environment is different than trying to be the best at something. We are talking about managing situations. It could involve *winning*, but this is not the primary purpose. In fact, a narrow Win/Lose mindset can hinder the process of Shaping our Environment.

On another night at Owl Farm, Hunter Thompson's fortified compound outside of Aspen in Woody Creek, I walked in and he was laughing and grinning from ear to ear as he wrapped a big green propane canister with nitroglycerine and duct tape. *Heh*, he chortled.

We went outside to the meadow in back and placed the target on top of an empty beer keg. Then he handed me a 12-gauge shotgun.

Growing up in south Louisiana, I learned how to shoot guns early on. My father's side of the family had a hunting camp on the Mississippi River at the intersection of Arkansas, Mississippi, and Louisiana. Hunter's farm was a much different environment. Cries of peacocks pierced the night. As I stared down the barrel at the homemade bomb in front of me, nerves shot through my body. *Did I really want to hit the target? Would shrapnel hit me and everyone else?* Hunter was crazy, but he wasn't careless. *Right?* But to be honest, my main fear wasn't being hurt, I was afraid of missing and looking like an idiot. I sighted the gun carefully, breathed softly, and fired.

BOOM!

The target exploded in a mushroom cloud. Sparks showered the sky. The flame reverberated and curled in on itself. Everyone hooped and hollered, me most of all. Back inside Hunter's kitchen, we talked some more about politics and the presidential election at the time. Hunter rambled on and then dropped another maxim, "*The reaction always defines the event, not the event itself.*"

Time has testified to the truth of Hunter's comment. What I remember most exquisitely about that night all those years ago isn't so much the explosion itself, but the reaction to it — the palpitations of my heart, the gasps and yelps from the small gallery of onlookers. The explosion was great, but the clarity of the moment right afterward is what sticks with me today. The relief of having successfully managed my nervousness and fear.

The *Rule of Reaction* touches on the ancient Stoic philosophy that we can't control the world, we can only control our impressions of it. Shakespeare's Hamlet puts forth an even bolder notion, "there is nothing either good or bad, but thinking makes it so." Milton's Satan apparently concurs: "The mind is its own place, and in itself can make a heaven of hell, a hell of heaven." That's not just him looking on the bright side after his fall from grace. Meaning is created when we *react* to a phenomenon and is shaped by the nature of that reaction.

The *Rule of Reaction* is a lesson I would learn time and again far from Hunter's kitchen. My professional experience has hammered home the importance of reaction.

How people react to something will determine its relevance and impact. No event ever happens in a vacuum. Again, sports illustrate this clearly. If you run the world's fastest 100 m, but no one's around to see it, how many gold medals do you win?

Dr. Cuddy crafted various experiments to put people in "Power" situations and then analyzed testosterone and cortisol levels. Confident people who feel powerful naturally exhibit high testosterone levels. Coupled with high cortisol levels, however, this creates a sort of madness that can drive someone over a cliff to arrogance. Dr. Cuddy found that people who exhibit the best nonverbal communication also have low cortisol levels. This is because they are adept at managing their reaction to stress and crisis situations. They keep a clear head. According to Dr. Cuddy, good leadership — real power — is not just about being confident, it's also about managing your reaction. Perhaps this offers a scientific explanation of the *Rule of Reaction*. In short, Dr. Cuddy shows how our internal story — what we believe about ourselves — is fundamental to what we become.

One of the main elements of crisis control is managing what happens next after the deal goes down. The event has already happened; what happens next tells the whole story. Communication scholars agree. The way people respond is much more important than what actually happened. In *Images that Work: Creating Successful Messages in Marketing and High Stakes Communications*, authors J. Roland Giardetti and John Oller write, "The report is more important than the event." The "report" — the news article, social media, blog post, or book — is not just what history will remember, but how it will be remembered. And while it may be simply one perspective, it becomes part of the historical record. The report is the analysis of what the events meant and the interpretation by certain actors with their own unique perspectives.

The lesson here is one of the most important for any person of responsibility. Whether you are managing a political campaign or your family, remember: it's about *Managing the Reaction*.

6

Friction

Expect the unexpected. Products don't move. Ads flop. Votes don't go as planned. Resources don't materialize. Coalitions fall apart. Someone somewhere says something outrageous and offensive.

But wait, we're just getting started...

Sponsors don't get along. Computers are hacked. Video conferences are zoom-bombed. I've been a part of events where the power goes out. A restaurant opening where the pipes explode. A nursing home catches on fire. The Russians hack your candidate's personal email server. Maybe it's something more personal: snow starts falling inexplicably in the middle of your 200-person backyard crawfish boil. Too many people show up or too few. They show up late or not at all. Caterers have the wrong menu. The band is drunk. It's always something.

"Friction" is a term coined by Clausewitz. He said, "Everything in war is very simple, but the simplest thing is difficult. The difficulties accumulate and end by producing a kind of *friction* that is inconceivable unless one has experienced war" (italics mine). The delta between planned results and actual outcomes indicates the amount of friction. Plans go sideways. Visions fail to translate. Consumers don't like the product. Other actors appear and derail your plans. You don't get the story you hoped for. People quit. Instead of praise, you receive ridicule and criticism.

Friction produces heat. The rub. The communication strategist's job is to manage this friction, to operate despite a certain inevitable amount of it. One must marshall those elements within one's control. We communications strategists are the Fire Department when friction gets so intense that it ignites. We are also the ones who get burned. Our role is not just to minimize *friction*, but also to absorb it, if need be. Like molecules, sometimes we bond and sometimes we collide. There is always the possibility — the

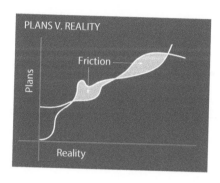

FIGURE 6.1
"Plans vs. Reality."

inevitability, to be honest — that shit will hit the fan. The challenge is how to manage it. If the job is done well, the wider public, or your boss, may not even know there was any friction at all.

The wise rabbi says, "Want to hear God laugh? Tell him your plans." Plans change because life tends not to conform to them. Therein lies the beauty. Life is live theater.

Our job is to minimize this delta between plans and the inevitable contingencies.

KRISIS

The term *crisis* comes from the ancient Greek word *krisis*, which is interpreted as a choice, a decision, or a critical turning point that will change one's destiny. The structure of Greek tragedies revolves around a moment in which the protagonist must make an all-important decision. According to the nature of such plays, these pivotal moments are more instructive as warnings than as models. After all, poor choices with life-or-death consequences make for more captivating drama.

All sorts of challenges present themselves during the production of an event, in a campaign — or in just about anything for that matter. A fair amount of a strategist's job consists of managing these challenges and keeping the plan moving forward. A good strategist knows when to shift and pivot and when to hold course. Anything can happen — and usually does. It's difficult to know what the threshold is for abandoning a plan because every situation is so different as is the amount of personal investment. It's not a matter of *if* the unexpected occurs, it's a matter of *when*. It's the Rule of Reaction all over again: our success depends on our *reactions* to these contingencies. The mission is to ensure that all the pieces are in place for success and then to be ready to adjust them to new circumstances as soon as they arise.

While the prescribed course of action is to be proactive rather than reactive, sometimes situations require a reaction. (Keep in mind, deciding not to act is also a reaction.) These are crisis situations. A crisis can strike at any time and usually does without warning. How a company or individual responds may mean the difference between success and failure. A bad response can make a bad situation worse. Much, much worse, in fact. Sometimes a crisis can be a costly nuisance, sometimes it's the thing that puts companies out of business for good or costs politicians their jobs. There can be no investigation of communication without *crisis*.

In a crisis, when there is a vacuum, the job of the communications strategist is to fill it with the most accurate, most relevant information available. If you don't, others will fill it with information that benefits them — that which they think may have happened or even *what they wish had happened*, regardless of its veracity. I've handled quite a bit of crisis communications and, to be honest, nearly everything has an element of crisis and urgency about it (at least, it should).

In *Crisis Management: Planning for the Inevitable,* author Steven Fink explains that crises are "merely characterized by a certain degree of risk and uncertainty." He points out that a crisis can be anywhere on the spectrum of good to bad and usually includes some of both. He refers to a crisis as "prodromal," a precursor of events to come. Fink writes,

> If any or all of these developments occur, the turning point most likely will take a turn for the worse... Therefore there is every reason to assume that if a situation runs the risk of escalating in intensity, that same situation — caught and dealt with in time — may not escalate. Instead, it may very well conveniently dissipate, be resolved.

In essence, a *prodromal* occurrence is the moment of choice.

Wouldn't it be so much simpler if all the events were essentially small plays and dramas that led audiences along prescribed emotional journeys? Unfortunately, you'll find that events rarely turn out this way, so you may need to put aside all of the notes from rehearsals, all the staging, and direction — you might have to throw out the script altogether! — and rely instead on your improvisational skills. We must maintain the capacity and the courage to change tack. It is this ability to self-analyze and re-evaluate that determines the fate of the strategist.

Traditional media, such as newspaper, radio, and network television, are just not as important as they once were, which is sad and disappointing. There is a paradox in news and with newspapers, television stations, and radio: everyday the headlines get worse, the stories more tragic. The news — and subsequently the world the news purports to represent — becomes more depressing. It often evokes despondency in its consumers. A feeling of hopelessness. Which then drives people away from reading news.

This cycle feeds the power of social media. People increasingly turn to be informed by their like-minded peers. It has also given rise to disinformation campaigns, election interference, and, yes, the Orwellian term, *Fake News*. (A U.S. Senator once told me that if I wanted to understand the world and what was going on, I should read the *New York Times* everyday.)

In recent years, social media has forced corporate America to deal more directly and more rapidly with racially charged PR catastrophes. U.S. society is a tinderbox and social media, by its profound powers of dissemination, adds accelerant to any conflagration. For instance, in February 2017, the public relations team for a large restaurant chain was in crisis mode after employees at a franchise restaurant in Missouri called the police on two black women whom they falsely accused of dining and dashing during a previous visit. The company fired all three employees involved before permanently closing the restaurant location, costing the company more than $21 million in damage and loss of revenue.

The world largest coffee franchise closed all of its stores for an afternoon of diversity training in the aftermath of a widely reported incident in which an employee called the police on two black men whose suspicious activity amounted to no more than sitting in the coffee shop. The total financial loss (in profits and brand damage) was more than $100 million.

Under similar circumstances, a massive clothing retailer fired three of its employees in West Des Moines, Iowa after they falsely accused a black patron, James Conley III, 29, of stealing the coat he was wearing, despite his explanation that he'd received it as a Christmas gift. (Surveillance video showed him wearing the coat into the store.)

A global pizza giant has faced similar struggles. The company's founder, John Schnatter, was forced to step down as CEO after criticizing the NFL's handling of players protesting police brutality during the national anthem and claiming that the protest hurt company's sales. He didn't like players kneeling. Schnatter's leadership role ended not long after the Daily Stormer, the white supremacist website, gave the franchise an official endorsement. The controversy forced the franchise to end its longtime sponsorship of the NFL because of declining sales.

And the list goes on... Denny's, American Airlines, Equifax — they all made boneheaded decisions that cost hundreds of millions and jeopardized lives and dignity.

On the flip side, it's also hard to come up with a lot of good examples because if they are handled well, we shouldn't hear about them.

Communication is so important precisely because the stakes can be so high for so many. Not only can it ruin reputations, it can cost people jobs and bankrupt companies.

A crisis can be handled the right way, which is fairly, decisively, and promptly or companies can continue to try to sweep them under the rug. At their own peril. A 2011 article about PR success stories in *Business Insider* advises, "Most importantly, companies that make mistakes must sincerely accept responsibility for their actions — not distance themselves from them."

The one thing that good responses to (potential) crises all have in common is that the companies acknowledged the situation and met it head-on. The CEO of the coffee company, learned this lesson the hard way during its racial crisis. He reflected, "The most important lesson we learned throughout our entire crisis came down to one thing: we should have acted sooner."

According to the 2007 article, "Reputation and Its Risks," in the Harvard Business Review, "Most companies, however, do an inadequate job of managing their reputations in general and the risks to their reputations in particular. They tend to focus their energies on handling the threats to their reputations that have already surfaced."

The ways to protect a brand in a crisis are fairly simple, but take courage and transparency. Accept and acknowledge there is a problem. People are hurting. The only way to resolve a problem is to accept it. Then figure out how to find a path forward that works for everyone. Finally, make it happen and strive toward a meaningful resolution. Through this process

RESPONDING IN A CRISIS

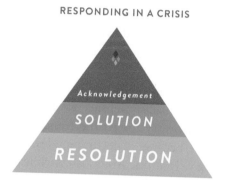

FIGURE 6.2
Responding in a crisis.

there should be some internal soul-searching about how to prevent it from happening in the future.

The point being, as a strategist you will be pulled into all sorts of situations, and understanding how to handle a crisis — or prevent one from happening — is a critical element of being a professional. Just like *Ghostbusters*, you want to be the one they call when the spirits are on the loose.

THE COURT OF PUBLIC OPINION

Let's delineate between the law and the public. While some issues involving real harm and unintentional damage are handled in courts of law, others are decided in The Court of Public Opinion, which has its own peculiar ways of rendering judgments. Abraham Lincoln was right: *He who controls public opinion controls the world*. While attorneys and 12-person juries handle matters in courts of law, another sort of jury — the public — may wield just as much influence (sometimes more) on the overall outcome of a legal case. Communication with the public needs to be managed toward favorable outcomes, but it's not always possible. After years of experience managing high-profile cases — those that involve murder, fraud, sexual harassment, abuse, and celebrities — I've concluded that sometimes you can have an impact on the outcome and other times managing the body blow is the best you can do.

I've worked with many law firms and have found that while some attorneys are masters of dealing with the media and getting their story out there or discrediting the other side, a fair share of them don't like to speak to the press at all for fear of saying something that may prove to be self-afflicting. This is a very real concern. Many attorneys are somewhere in between, depending upon the context. Still, leaving an information vacuum can be even more damaging. A good communications professional fills this void. This can mean the difference between the public perception of being guilty or not guilty. In law, as in much of life, the person who tells the better story usually wins.

One of the age-old questions is *Do judges and the legal system adhere strictly to the letter of the law or can they be swayed by public opinion?* It depends on the particular judge. My friend and communications expert, Curtis Robinson says, "It can't be both." If judges are repelled by the media, then they're influenced. They're not unbiased observers. My wife, a top water rights attorney, concurs.

Gerald Goldstein is former head of the National Association of Criminal Defense Attorneys and one of the top criminal defense attorneys in America. Because of the nature of his high-profile clients, he is a wizard at knowing when to use the media — and when not to engage. He's been in front of the press countless times on justice issues. He was also an attorney for Hunter S. Thompson, which is how we became friends. Before working with Gerry on a number of high-profile cases, I expected him to have an aggressive posture on using the media, to be the public bulldog, but I was surprised by his hesitancy.

He contends that a big challenge for an attorney is not having ultimate control of the narrative. When attorneys speak to a journalist, they must understand that the journalist has the last word. "Don't kid yourself," he says. "They are the ones writing the article. It's going to be in their words and you may not agree with it."

Gerry notes that the courtroom is a much safer place. When attorneys bring a case to court in front of a judge and a jury, they have control. To engage the public, attorneys can communicate through court documents, which the media can also use, if they are interested. Gerry says, "Good attorneys write their briefs for the newspapers and the public, not the jury and judge." He points out that judges rarely read briefs. "But they do read the newspapers." Venture outside the brief and you're entering a danger zone.

The problem is that prosecutors and government attorneys usually have a cozy relationship with the press. "The media sup at that trough," Gerry says. "They love an easy source." Prosecutors and District Attorneys can feed information to the media and if you, the defense attorney, fail to respond, the resulting information vacuum can hurt your client.

But Gerry cautions that you can only influence the court of public opinion if you're in the right. The events of the case have more to do with the media outcome than the attorney does. "Just because public sentiment is with you doesn't mean courts will go along," he says. Public opinion can give the courts cover. Judges generally like to know that what they are doing is for the common good and not bucking public opinion. Gerry avows that it's all about the truth, facts, and justice, and unless these are on your side, there is no way to influence public opinion, much less judges and juries.

"Don't think for a minute that what you're doing is going to make a difference," he warns. "Unless the Truth is on your side."

7

State of the World? Desperate as Usual

Everybody has a plan until they get punched in the mouth.

Mike Tyson

The jabs, hooks, crosses, and uppercuts Mike Tyson endured and inflicted are nothing compared to the toll of a global pandemic, which is no ordinary disruption or crisis. Not now. Nor in 1918 during the "Spanish Flu" influenza epidemic, which, as it's been estimated, took the lives of over 50 million people. While many companies and organizations have crisis plans for an active shooter, a financial free fall, fraud, or an extreme weather event, few have prepared for a global crisis of a deadly infectious disease. In uncharted waters — pandemics, climate change, growing economic disparity, racial tension, rising geopolitical hostilities, and everything else — the fundamentals of a good proactive communications are more important than ever.

As communications professionals and leaders of organizations, companies, and agencies, you will be faced with how to communicate in uncertain and chaotic times. Even though the COVID-19 pandemic is a unique situation and the stakes are much higher, it's important to remember the best practices and protocol in managing any crisis.

Even if the world is literally closing down, your communications, both internal and external, should not. In fact, this is the time to ensure that the way you talk about your organization is especially straightforward. Perhaps this is your chance to shine. As with any crisis, there is usually an opportunity lurking in the shadows.

When managing any fragile issue, such as a pandemic, the most important element and guiding principle is to maintain **Trust**. Remember that the communications business is about getting people to care. Some degree of trust is always a prerequisite of caring. People must trust that a company's leadership is living up to the stated values of the organization and ensuring the health and well-being of those they are charged to protect. Good leaders always consider the people who make up their world and depend upon for success, including customers, members, employees, regulators, and investors. Maybe your company is doing many positive things, but if your own stakeholders don't know about it, then your actions might not make a difference to those people who matter most.

Trust can be simply understood as the consequence of a promise fulfilled. This includes both explicit promises — like reliable internet service, consistent power, and water delivery — or implicit, such as living up to brand identity. Are expectations being met? Trust is broken when people believe you have broken promises or not lived up to expectations. Trust can get tricky because sometimes expectations are defined by us, but many times expectations are set by others and are beyond our control.

Expectations can be a murky, dangerous swamp.

The first rule of a crisis is understanding that trust is built and maintained by living up to these stated values and expectations. Despite the best laid plans, events happen, other actors appear, the council vote gets cancelled, the state capitol closes. Good leaders keep marching through the crisis and transforming themselves in the process. However, this isn't what matters most. What is most important in considering how to react is determining what people *think* you should do.

You begin a successful response by putting yourself into the position of others by understanding their expectations. How do those interests *expect* you to act? What kind of action do they call for? Good leaders understand that a response is not about personal preference, but what others expect of them in certain situations. The good news is that it is possible, to a certain extent, to map those interests.

We must remember that in a crisis people often feel incredibly afraid and therefore vulnerable, both professionally and personally. In fact, the hallmark of working with people in the middle of an intense situation is that it becomes very personal very fast. You never really know what someone is going through in their life, but our job is to rise above the inevitable internal tension to look at the entire worldview.

Which means that trust and empathy in any communication are more important than ever.

We expect our leaders to *care*. In any crisis response we must act and show people that we care. We cannot just *tell* them we care, but must *show* them that we care through our actions. The single biggest mistake in a crisis is indifference. Only by a consistent demonstration of empathy can we maintain confidence during a crisis. Any crisis response strategy must begin with a declaration of our values, which explains *why* we care.

Many in leadership and in corporate suites may feel a tendency to hold off from communicating *until they know more*. (But that day may never come!) Lawyers are particularly skillful at avoiding public attention. As I mentioned in the last chapter, oftentimes lawyers discourage saying anything in public because public statements do entail some risk. Sometimes attorneys will write something that only makes sense to a judge and avoids the Court of Public Opinion altogether. In fact, we frequently have a push-pull relationship with attorneys when it comes to when and how to respond. This can be healthy. (Nothing against lawyers...I'm married to one!) In a big crisis there will always be lawyers who have their own style of language, usually watered-down mush that no one understands. I've learned the hard way to say something like, *Well, I'm not a lawyer, and while your statement might make sense from a legal perspective, it'll seem like our head is way up our ass and that somebody is going to jail.* Our job is to take legal jargon and turn it into language that is clear, concise, and usable. Attorneys in particular bristle at being told how to write or what to say, but this is where we bring value.

When on terra incognita, leaders must demonstrate how they care and they must show empathy. Arrogance and denial make empathy impossible. In times of uncertainty, the clarity of direct language is paramount. Avoid legalese. Don't dance around an issue, be honest with your employees, customers, and stakeholders. Don't use euphemisms or half-truths. Don't say, "restructuring," when half the company is getting fired. Don't say you won't lay off people when there's a chance you might have to.

Not communicating at all, however, is counterproductive. Silence is interpreted as indifference, which is *anti*-empathy. *Why aren't we hearing anything?* invites opportunists to join the conversation. Remember, if we aren't defining ourselves, then others will do it for us. In a crisis, there is a maxim that the longer the silence continues, the less control we have over the outcome.

Perhaps one of the biggest questions confronting (and confounding) leadership is: *How do we know when it's time to take action in a crisis?*

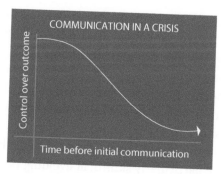

FIGURE 7.1
"Response and Time in a Crisis."

Helio Fred Garcia, author of *The Agony of Decision: Mental Readiness and Leadership in a Crisis,* identifies four criteria for determining whether we should take action in a crisis:

> *Will those who matter to us expect us to do something now?*
> *Will silence be seen as not caring or as affirmation of some kind of guilt?*
> *Are others speaking about us and shaping the opinions of those who matter to us?*
> *If we wait, do we lose the ability to determine the outcome?*

If the answer to any of these questions is *Yes,* then it's time to act. Assemble the crisis team and determine who will be in the bunker calling the shots.

A global pandemic is precisely the time to ramp up communication to keep the brand and reputation strong during times of uncertainty. Your organization should show leadership by becoming part of the solution. Assuming full compliance with suggested safety measures, what can your people do to help? Maybe preparing lunches for underprivileged students, delivering meals on wheels, or offering online courses and professional development. Keep communicating and being transparent. Stay relevant to your stakeholders' lives. Keep your company active and engaged with your clients, vendors, and stakeholders. Keep assessing both best-and worst-case scenarios, and create communications plans accordingly.

As in most bad situations, chaos, pain, and challenge usually bring some kind of opportunity or silver lining. *What can these hardships teach us?* Possibly to show more gratitude for the blessings in life? To appreciate colleagues more? Think about using the opportunity to practice mindfulness and a keener sense of situational awareness. To bring a sharper focus to your work. A crisis can be a good time to reset and recalibrate. There is nothing like a crisis to identify what's really important by filtering out all else.

In my experience, during times of uncertainty and upheaval communications is always the first to fly out the window. During the 2020 pandemic, many clients cut back staff and resources. This is precisely the wrong move. Good leaders know that enhanced communications are even more important than ever. These are the *prodromal* moments in which brands and organizations live or die.

Remember, empathy and trust should be cultivated in advance of any crisis, and not just because they constitute the strongest bulwark against devastating prodromal events. In actuality, the crisis plan should be nothing more than an extension of the everyday operating procedure, part of our natural daily caring, a piece of that rock we're always pushing up the hill. Our work is never done.

These are universal truths in how to think about being in crisis situations. As historian Carol Byerly points out in her book, *Fever of War*, about the 1918 Influenza Pandemic, massive cultural and economic shifts have happened before and are certain to happen again. In a sense, this is the Drama of Humanity. Those who communicate and collaborate well tend to survive more than those who don't.

Communication Darwinism.

The job of good leadership is to convey a positive, practical attitude, to constantly assess, to remain forward-looking in difficult times, and to prevent panic. Most importantly, don't forget to take care of yourself. As a crisis communicator, you can't afford to succumb to the same fear and panic that you are trying to alleviate in others. While it feels like the world has been punched in the face like legendary boxer Mike Tyson, we must stand tall and keep on swinging.

SINGLE OVERARCHING COMMUNICATIONS OUTCOME

In communicating during times of a public health crisis, we need look no further for a fine example than the World Health Organization. WHO has produced an *Effective Communications: Participant Handbook*, which is also applicable to any communications regarding public health — or any policy issue at all, for that matter. The handbook states, "The most fundamental skill that a good communicator possesses is a clear understanding of the change they want to see regardless of what they say or how they say it." To do this, we must first develop a **S**ingle **O**verarching **C**ommunications **O**utcome (**SOCO**) and get to The Point fast. Keep in mind, this is coming from an organization that understands that if they don't explain matters properly, people will die.

The SOCO is the change you want to see in your audience as a result of your communication. SOCO is an outcome, and must therefore be expressed from the perspective of the audience. It is *not* an objective, which usually reflects a perspective or values. The SOCO must be explicit about the change you want, and it must be time-limited. WHO asserts that the SOCO must be realistic and achievable. It must contribute to a larger program goal or objective. The SOCO framework applies not to just public health issues, but any high-stakes communications.

There is a framework for how to develop a SOCO.

Step 1: What is the issue?

Step 2: Why do you want to focus on this issue, and why now?

Step 3: Who needs to change their behavior?

Step 4: What is the change that you want to see in your audience as a result of your communication? (THIS IS YOUR **SOCO**.)

WHO applies this framework to pandemics, outbreaks, antimicrobial resistance, and every other health crisis. But it can be applied to any communications situation.

Remember that people do not listen or hear in the same way. We listen best when our attention is captured and focus is achieved. Once we are interested, we will take the time and make the effort to listen to the nuances of an argument or explanation. This becomes even more important in a world transformed by communications technology. Listeners, including experts listening to other experts, are inundated with competing information. We need to get to our point as soon as possible and explain the supporting information with all of its requisite subtlety and shading only incrementally and in decreasing order of relevance and importance to our audience. For all essential communications, model your speech on how listeners actually take in information. Transcend

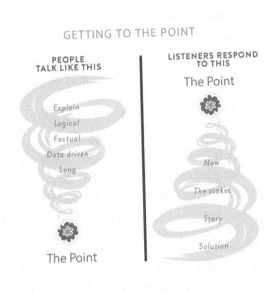

FIGURE 7.2
"Getting to the point."

party divisions and base information on science — that is what will give people confidence. WHO contends that in a pandemic it is critical to prepare people for the future. What do we tell people today to get them ready for four days from now? Solving for the future.

DEATH OF TRUTH

H.G. Wells noted, "Human history becomes more and more a race between education and catastrophe." Leadership is difficult without trust, empathy, and updated information. While these characteristics need to be enhanced and heightened during times of crisis, pandemics, hurricanes, and floods, they are also the keys to stronger communications, day in and day out, both professionally and personally. Yet we keep learning the same lessons again and again, repeating the same mistakes. Trust seems to be in short supply. A lack of trust that our leaders have our backs. That we are in this grand endeavor together, no matter what business or political party.

As our pandemic historian Dr. Byerly points out, diseases, pandemics, floods, fires, have always marked the course of humanity. "Well before

Copernicus and Galileo, we have been in a constant struggle of science versus religion versus politics," Byerly tells me.

> With the virus, we are not in charge. Unless there is a vaccine, we are not in control. In a pandemic, keeping the public healthy all boils down to clear communication and good information. But our bigger problem is that many people no longer believe in experts or scientists.

Dr. Carol explains how in the military there can be *moral injury* when soldiers don't trust their commander. They can even develop PTSD if they believe they are fighting an unjust war and that they are being sent to die in someone else's battle. "Society can have a moral injury if they think their leaders don't care about them. People have to believe in their government even if they don't always agree with their leaders," she said. "The single biggest factor in a government is legitimacy."

"Today we are shell shocked and battle fatigued from misinformation," Dr. Carol, my neighbor, laments. One afternoon, on her front porch, as a storm moved in, she elaborated,

> Our world has changed. After WWI, during the 1920s, radio created a national voice. Think of FDR's Fireside Chats.
>
> It seems like we're now seeing the Death of Trust. It actually started with the Vietnam War and Watergate. The difference then was that cynicism was about the government. Now the cynicism is about expertise. And that is frightening to me.

"Who wears a white coat anymore?" she asks. "Nobody." Without a shared set of facts, how can we govern? Dr. Carol is careful to qualify that this is not happening now all at once. Rather, it's an evolution of our American story. "Epidemics always show what a society is made of and what we are seeing now is exposing the lack of sinew that holds us together."

Satya is the Sanskrit word for truth. It can be interpreted in two different ways. The universally agreed-upon truths ($2+2=4$, Entropy, etc.), but the other more important interpretation is about the virtue of truthfulness in a person. The qualities of honesty, trustworthiness, and integrity. It is this second notion that is most under assault by those who seek to destroy the truth. Who protect themselves at the expense of the common good? What the COVID pandemic laid bare, what all disasters and tragedies expose, is not the naked dead or shattered economies, is the true peril that truth is now facing. It has shown the incapacity of governments to protect their people by telling them the truth. Not just in 2020, but with every disaster, there is a new era of "Alternative Facts." Disputing science with rumor. Our criminal defense attorney Gerald Goldstein also laments the "Death of Truth by a thousand cuts." But he also believes, ultimately, "Truth will rule the day."

In a quote widely attributed to Senator Hiram Johnson, "The first casualty when war comes is truth." While I appreciate a good story, the best is rooted in truth, not deception. As we know by studying history, a civil society cannot function in a noxious cloud of distrust and self-dealing. Hurricanes in New Orleans and Puerto Rico, the oil spills of the Exxon Valdez and British Petroleum in the Gulf of Mexico (whose CEO famously wanted to get back to his golf game), SARS, wildfires in the West… We have experienced great tragedy and disease, and we will again.

"These incidents are not so different, yet we keep learning the same lessons over and over again," Dr. Carol says. "The stakes are high in getting the communications right. Not just to sound good, or even because it can be costly, but because when they don't believe the science, when a crisis is about politics, *then people die.*"

8

The Red Baron Meets the Buddha

Imagine your mental focus as a Maglite. You can adjust the flashlight beam so that it's super tight or very wide. Let's do away with the clutter, bear down, and go tight. Concentrate on a single point 6–8 feet in front of you. As you breathe, bring your attention to the point at the end of your mental beam. Stay laser-focused on that point for seven breaths. This is called **Riding the Breath**. When you get distracted — and you will! — start over. Do three sets of seven breaths.

When life gets tough — stressful client situations, long-distance swims, tight deadlines — I rely on mental conditioning techniques such as *Ride the Breath*, described above. These practices help allay stress by recalibrating the nervous system, but they also help sharpen focus and increase awareness.

Here's another. This one is called *The Zone*.

Instead of concentrating on a single point, open the focus on your Maglite and take your awareness to the widest possible angle. As you breathe, become aware of all of your senses — smell, taste, touch, sound, sight — with the sensitivity of a hunter stalking its prey. Discover where you are. Find your exact place in the world. Do another three sets of seven breaths.

Now practice alternating sets of Riding the Breath with sets of The Zone. Tight to wide and back again. To build the muscles of attention and awareness, practice shifting the Maglite of your mind between being hyper-focused on a point and being fully expanded, taking in every aspect of the space around you.

I've learned these exercises from an expert in Jedi mind tricks and mental conditioning. Mark Williams is a former F-16 fighter pilot who's seen his share of combat. Like Tom Cruise in *Top Gun*, he's handsome, a smart ass, and extremely good at what he does. Unlike Maverick, Mark has run for Congress and developed a seven-hour "Learn to Meditate" course on the intricacies of *Riding the Breath* and more. If that wasn't enough, Mark has drawn on the mechanisms he learned to focus his attention and augment his awareness when flying jets to develop a mental conditioning practice for top athletes, CEOs, military, first responders, and anyone else who wants to be prepared and keep their composure when things go awry. Mark's special brand of mental conditioning is a particularly useful tool for the strategist seeking to develop and maintain situational awareness (SA) and establish clear lines of communication among team members.

I've known Mark for a long time. We first met working on a congressional race where he was our foreign policy and military expert. After flying in the military, enduring personal tragedy, and kicking the tires and lighting the fires of civilian life, Mark enrolled at Naropa University, where he developed his mental conditioning practice. He wrote about the practice in a fantastic article called, "The Red Baron Meets the Buddha." Incidentally, he was also my support paddler on two of my long swims. He paddled a kayak in turbulent waters for 16 hours as I swam across the Caribbean. He also stuck by my side when I made the first-ever swimming descent down the Colorado River through Canyonlands for 47 miles in just over 14 hours. Mark is unwavering in the face of difficult circumstances, to say the least. He is the person you want on your team in tough spots.

Mark is a great teacher of situational awareness because in his flying days, his whole life depended on it. One wrong move, one misreading of his environment, and that was it... No second chances. As he explains, speaking from personal experience, when a Russian MiG fighter pilot is on your tail trying to shoot you down and your own plane is turning aggressively this way and that, it's difficult to keep track of your buddies or other bandits who can make or break your day. That's why fighter pilots spend so much training time developing SA.

With SA, Mark knows the composition of space, the movement of enemy fighters and their relation to friendlies, who has the advantage or disadvantage, and most importantly, whether or not to strike.

"Without it," he says succinctly, "you will die."

Mark and I were on a four-day river trip through Canyonlands on the Green River to scout another long-distance swim and had a chance to discuss a lot of ideas. We talked about situational awareness, group dynamics, the importance of planning, and other similarities between military aviation and communications strategy. He suggested that success in either vocation — success of almost any kind, for that matter — could be attributed to the same thing.

"Being a self-aware social creature," he said,

> brings the capacity to understand what's happening inside yourself and also the environment outside without saying a word. Be brutally aware of your own game, your level of craft. What do you do well? Where do you struggle? Where does your client do well? Where do they fall apart? The more you can be aware — not just from the left brain, but from a right-brain/emotional intelligence perspective, the better off you will be. The cleaner and clearer your mind, the better you see. Like a mirror.

The mirror metaphor reminded me of Hamlet as an acting coach. I wanted to hear more, and Mark indulged me.

> The best mirror is the one with the cleanest reflection and no distortions. If you are a mirror with distortions, then you are not seeing so-called 'reality' clearly. You can't orient well if your mind, your mirror, is dirty with ego or distorted with fear. The more you can polish that mirror — by knowing yourself, removing your internal detritus, establishing relationships, placing trust in other people, and developing real connections — the clearer you will be.

And here's where Mark made a very interesting shift to situational awareness. He conceived of it as something broader than the layperson might expect. SA is not confined to a sensory understanding of one's immediate surroundings — getting the lay of the land, so to speak. It pertains as much to understanding those involved in any act of communicating. Making and maintaining real human connections, in other words. Empathy, again.

I asked Mark, *What role does communication play for a fighter pilot?*

He thought for a moment, and I could see his mind working. "The way pilots communicate is the ultimate test of their tactical strength," he said.

> The pressure is extreme. Flying along at 500 mph, things are happening quickly; add in bandit aircraft trying to get a shot off, SAMs (surface to air missiles) and AAA (anti-aircraft artillery), now life can get extremely spicy. If you fail to communicate correctly, or everyone is talking on the radio at the same time, no one understands a thing.

Mark was getting charged up and explained the discipline of tactical communications in combat and how fundamental it is to fighter pilot training.

As silly as it might sound, when they say something like *Blue Angel Check*, there is a precision to the communication that's being taught on Day One of pilot training. It is extremely serious. You have a lot of people who want to share a lot of information, but you can't do that. You learn to communicate succinctly, precisely, and only as needed. If you mess up, someone could die.

Fighter pilots in training that fail to master the "comms" are ignominiously "washed out" of the schoolhouse. "Them's the stakes."

Mark cautioned, "*Never shout.* Fear is transmitted over the airwaves, and it spreads like a pandemic. When there is fear in the air and blood in the water, then the enemy comes to feed. Always stay cool." He said the same rings true for anyone engaged in a crisis or confronting a big decision.

One night around a campfire on a big sandy beach on the river, he referred to a YouTube clip of four F-16s coming to drop the first bombs on a nuclear reactor in downtown Baghdad. Surface-to-air missiles are everywhere, but the cadence of the comms is right out of a textbook: smooth and surgical. Mark said,

Even if your buddy gets speared by a SAM, nobody freaks out. Everyone just moves on and executes the plan until it's wheels down and safe at homebase. Then the debrief, shots of whiskey to honor the fallen, and finally the grieving begins.

Mark explained that before mission launch, everyone gets briefed on the communications plan. I asked what a communications plan even looks like for a fighter pilot. There is a standard brief: *This is how the mission goes.* Everyone knows who is in charge of what and their distinct role. The objectives are clearly spelled out. The Comms Plan is "standard" — we understand the pace of how we communicate when and why and what the main messages are. "When we create a plan, we are building an ideal picture of our world, in essence telling the story," Mark says. "But since life is never 'standard,' especially in combat, we always create contingency plans for when things go in south — and they usually do."

In this way, there isn't much difference between fighter pilots and communications strategists. Success begins with a standard playbook. Everyone must know *This is how we execute.* It must be part of your work culture's DNA. Whenever you bring a newbie on board, whether formally or informally, you must give the repeatable explanation, *This is the playbook — this is what our world looks like, this is how we operate.* You can tweak a bit, but for the most part, you should follow the procedures.

Heightened awareness confers success on all kinds of relationships, not just among team members. It is also vital in handling problem clients. Mark observed,

> If you have the good fortune of sitting down with a client ahead of time and you can understand and be aligned and in sync, then you have a much better chance at success. You have to have the mind meld before the crisis moment. Getting as aligned as possible on their goals.

I wrote in my journal Mark's quote from Sun Tzu,

> It is said that if you know your enemies and know yourself, you will not be put at risk even in a hundred battles. If you only know yourself, but not your opponent, you may win or may lose. If you know neither yourself nor your enemy, you will always endanger yourself.

Mark also referenced decision making, our next chapter, which he had studied in the military.

He said if the situation is more than just you and you're Blue One with the rest of your team in formation, then you have to have clear, crisp communications to build the Situational Awareness that leads to winning decisions and actions. If your enemies make better decisions faster and more efficiently, they likely win. If you want to destroy a target, isolate them, cut off their communications, and cripple their SA.

"Then they're fucked," Mark said.

9

Decision Making

So how does one make decisions in a crisis — or any situation for that matter? Fortunately, there is a simple framework for how to respond when someone figuratively gets punched in the face.

The OODA Loop (Observe, Orient, Decide, Act) was created by Air Force Colonel John Boyd, a military strategist who used it for combat operations. Although he never wrote a book about the process, his OODA Loop was held in high regard within the military. In 1976, he published a crisp paper about the decision-making process called "Destruction and Creation." He wrote, "The activity is dialectic in nature generating both disorder and order that emerges as a changing and expanding universe of mental concepts matched to a changing and expanding universe of observed reality." Boyd showed how, "To comprehend and cope with our environment we develop mental patterns or concepts of meaning." He argued that we destroy and create these patterns to permit us to both shape and be shaped by a changing environment. We can't avoid this kind of destruction of our ingrained patterns and creation of new ways of thinking if we are to survive on our own terms. If we are to learn and grow. His overarching goal was "to improve our capacity for independent action."

Col. Boyd viewed decision making as a continuous loop — an engine — where over and over again, through Observing, Orienting, Deciding, and Acting, we repeat this cycle of *destruction* and *creation* until we demonstrate internal consistency, sync up with reality, and achieve SA. Of course, this never happens — except maybe in fleeting moments. Achieving and remaining in this state is purely aspirational. In actuality, developing SA so we can make the best decisions is a never-ending process. Trial and error and error and error. More art than science. We are forever blowing up preconceived notions and destroying concepts in order to integrate

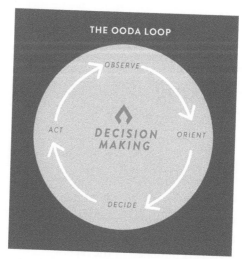

FIGURE 9.1
"The OODA Loop."

new information, relationships, and skills, which create new concepts of meaning and interpretations of reality, a new level of SA and hopefully better decision making. Col. Boyd's process is as relevant and true today in organizations as it is on the battlefield.

Interestingly enough, like the astrophysicist Dr. Jeff Bennett, Col. Boyd was also interested in the Heisenberg Uncertainty Principle and Entropy. He wrote,

> Taken together, the entropy notion associated with the Second Law of Thermodynamics and the basic goal of individuals and societies seem to work in dialectic harmony driving and regulating the destructive/creative, or deductive/inductive, action — that we have described herein as a dialectic engine. As indicated earlier, these mental concepts are employed as decision models by individuals and societies for determining and monitoring actions needed to cope with their environment — or to improve their capacity for independent action.

The OODA Loop is simple, but profound: when the deal goes down, what process should guide our decisions?

The OODA Loop looks like this:

Observe: A situational assessment. What is happening here? It seems easy enough, but sometimes you don't know all the players or have all the facts. Sometimes we don't know what consumers are feeling (despite being asked) or what voters believe (because they may not know themselves). Observation does not happen by accident. It is a skill which can be enhanced through training and practice. To observe is to be aware and consider the many levels and layers of complex situations, including the internal layers that bias your seeing. As our fighter pilot Mark Williams was keenly aware of: the cleaner the lens, the better the observation.

Communications professionals advising in a crisis need to use a tactical and surgical approach to understanding their environment. Only when you understand your environment can you begin to shape it and advise your client or others on how to interact with the unfolding reality. Observation at a high 360° level gives one heightened awareness with which one can make tactical decisions to achieve the desired strategic outcomes.

Orient: Once you have the lay of the land, this is where you consider your options. Start answering the questions: *What is my approach? What are the possible scenarios? What are my resources? What is on the line? What does success and failure look like? What and Who are affected by these decisions? What interests are at stake?*

Decide: Life is about choices. Find the option that is most appealing under a rational cost-benefit analysis blended with context and weighted against emotions. Most likely we go with what *feels* like it will produce the most favorable outcome. Out of all the scenarios envisioned, this is the decision point about where you need to go — both on the micro and macro level. Then there's all the planning decisions. What is our plan of action? Map it out. Strategy is what you intend to do and tactics is how you get there. (See Ignition Communication Template at the end of Part 2.)

We make our choices, but in the end, those choices make us.

Act: Let's get it done. Move. Make the thoughts and ideas happen. Let the tactics come alive. Remember what Harold Lasswell said, reality is created only when we make it so. But success is never free and plans never, ever, materialize precisely how you envision them, and if they do, consider it a rare unicorn moment. Even with all tactics implemented, there will always be the element of friction — players act unpredictably, team members underperform, targets keep moving, new information emerges. A professional's goal is to manage and minimize this friction when the *action* goes down.

Then start over.

Create and Destroy.

This decision-making framework is called a "loop" because as you act, the situation develops, which means elements change, new coalitions form, new resources become available, old resources dwindle, etc. You need to continually assess, and each assessment calls for another round of OODA. Old concepts are broken down or reframed and new concepts and meanings emerge from those ashes. You observe *again*, reorient, react so that you can keep determining which subsequent decisions/actions need to be made and when. It is an improvisational dance. The OODA Loop, which originated in reference to the military battlefield, can just as easily be applied to any crisis situation, however big, however small.

This model is reflected in Robert Fritz's book *The Path of Least Resistance*, in which Fritz describes the creative process as a cycle of Germination, Assimilation, and Achievement. These refer to the beginnings of an idea and the peak of its emotional appeal to the creator, to the harnessing and management of one's inner resources, and to getting it done. Key to Fritz's model is the notion that the shortest path to a goal is not a straight line, but the one best facilitated by the underlying structures of an endeavor. For instance, a river doesn't move directly from source to mouth, but follows a winding course along whichever way is easiest for the water to move. Furthermore, the landscape is dynamic. Water erodes the banks, a new path of least resistance is created, and the course of the river changes. But through iterative processes like the OODA Loop, it's possible to successfully intervene in a complex system. The river can be diverted or dammed — the structures of the system can be changed. We can tackle a problem from one mindset or try to alter the perspective forming our approach. Continue to observe, *reorient*, decide again, and act once more.

Over time, the OODA loop has been applied to many disciplines beyond the battlefield, including business. See Chet Richards's *Certain to Win*, a title that may have drawn a smirk from the notoriously irascible Colonel Boyd, who recognized, by the dynamic nature of the Loop itself, that nothing is certain in this ever-changing world. As Donald Rumsfeld, the former Secretary of Defense who led America into the war with Iraq, would say, there are known knowns and known unknowns, but there are also

4-STROKE ENGINE

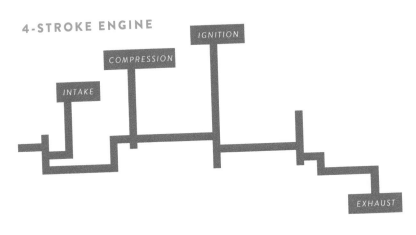

FIGURE 9.2
"The OODA Engine."

unknown unknowns, which are certain to keep your loop spinning and spinning.

The moment you think certainty is at hand could be the very moment when the world is about to be turned upside down.

Just like a pandemic.

It is worth noting noted the similarities between the OODA Loop and a Four-Stroke Engine:

Intake: pulls fuel in (Observing, Research)
Compression: Fuel is readied for spark (Messaging, Crystallization)
Ignition: The fire, when fuel is converted to energy; The source of power
　(Energy, Action)
Combustion: What happens to that power (Result, Reaction)

IMPROVISATION

The OODA Loop essentially describes making choices, which has a lot in common with the art of improvisation. In the face of an uncertain and unscripted future, improvisation is a mode of decision making that is instantaneous and taps into emotion as much as it does reason. A spontaneously evolving story. We may think we know how something will turn out, but then life throws us a curveball. Communication is the public theater of *Improvisation*. Shaping events, eliciting particular reactions, changing minds, making decisions, responding in crises, divining the right words to fit the emotional tenor of the moment, asking *What if...?* Both the communications professional and the improvisational actor deal in the moment. There is no script.

The legendary actor Bill Murray once said funeral, "Improvisation is the most important group work since they built the pyramids." Which might be why most of the cast of *Saturday Night Live* comes up through the ranks of improv troupes, such as Second City in Chicago. Murray was referring to the solutions attained when people work together and support each other. As a training tool for clients we sometimes bring in a professional improvisational coach. A close friend of our family, Alissa Ahlberg, teaches us and the people we work with to look to the future and consider the ways we can shape it by creating scenarios that heretofore did not exist. She explains how the principles that guide her interactions with others on stage also apply to business meetings and events. Alissa teaches that an important negotiation is a perfect example of an improvisational experience. A play of personalities, agendas, context, timing, and the spirit of the moment.

The very art of a communications professional is to act in an unknown, undefined moment. Our audience may not be those in theater seats, but they are investors, regulators, policymakers, journalists, wags, neighbors, parents, and a cast of characters who can choose to enter or exit the Drama of Humanity at any moment. It's a fascinating and oftentimes thrilling experience. It can also be frustrating and dizzying. But at the right moment, you know your words made a difference — both on the improv stage and in the boardroom.

At the end of a long Friday, Alissa and I sat down for a chat at our dining room table before she headed off to "Denver's Next Improv Star" contest. I wanted to know, *How does improv relate to different professions and to communications in general?*

She thought about it for a bit and said that it's not so much about any one profession; improv relates to life overall. The skills we use as *improvisers* are applicable to any endeavor.

One of the well-known first rules of improv is *"Yes, and..."* The players on the stage should not negate each other. She pointed out how in public relations, we are hired by a client to bring their ideas to fruition. We have to give them what they want, which is to take their position, grow it, and shape it. "You have to accept and then build upon the concepts," Alissa counsels.

Another important improv rule is more like a nudge: *Hey, it isn't always about you.* Sometimes we are our own worst enemies, especially if we steal spotlights intended for others. "Get yourself out of the way," Alissa says.

> The ultimate test for an improvisational actor is that it isn't about you. It's about the other person on the stage. Put your attention on the other person and make them look good. In turn, you look good. I should be able to go onstage with Amy Poehler (SNL, *Parks and Rec*) and make people clap for her. Not shooting down ideas. If someone gives you a bad idea, then it's up to you to make it brilliant. *You have to leave your own ego at the door.*

Alissa tells us that the skill of improv is about "Listening, connecting, and responding... with honesty." She pointed out the difference between "listening to understand versus listening to respond" and noted that it's not easy to forego your own agenda in order to be in the moment and allow conversation to happen organically, even if you know it's for the greater good. Del Close, the father of improvisational theater, advised, "Don't bring a cathedral on stage. Bring a brick, and let's build it together." I thought, *this sounds a lot like some people we encounter in meetings — so devoted to their ideas and preconceived notions, they consider them sacred temples.*

Alissa elaborates, "In improv we are in a performance, which is not always life. But what is always constant is that the ability to connect with someone is the greatest gift you can give any colleague or client." One of the most important lessons that improv can teach us as professionals — and simply as people — is that when you listen to truly understand, you open yourself up for collaboration and pass on the greatest gift: Real Human Connection. As Alissa rhapsodizes, "When you feel heard, it's the best feeling in the world."

She explained that although it may look like the actors are just making it all up on stage, they are actually following a set of rules. "Improv gives us structure and a framework to be efficient with our creativity," she said. Improv came into being as an artform when it was first used as a tool of sketch writing. The artists study structure and a format.

Improv works like this: Every time the *improviser* speaks or there is action in a scene, there is usually an *offer.* The actor throws out something that defines some element of the reality of the scene. These offers could include a location, name, relationship, or a pantomime to define the physical environment. These offers are known as *endowment.* It is the responsibility of the improviser to accept offers their fellow performers make. Refusing or failing to accept and doing something else is known as *blocking,* which is mostly frowned upon by other actors. (Just like in a meeting when someone blocks you.) When an improviser accepts the offer, they usually accompany it by adding a new offer, skillfully building off earlier offers. This is the process that results in "Yes, And..." It is the cornerstone of improvisational technique — and also of good teamwork on any professional project. The exciting part about improv is combining the unknown with the known. "We don't know what the story is going to be, but we know how to tell a story," Alissa says.

She shared her big question, both in her personal life and in her improv work: *How do I support someone, even if I don't really like their idea?*

This question blew me away. It's a struggle we all face — in professional and personal relationships — but seldom acknowledge.

Alissa talked about how our job is to rise above an average or bad situation to make it better. A lot of this consists of listening to what someone says. If you are listening and lending support for the greater good — then it's not about you. She said we must let go of our agenda and biases and be a part of something that is greater than the sum of the individual players on the stage — it's a powerful lesson and vital in creating a frame for how we communicate, respond, and react.

The courtroom, the daily newspaper, the influencer video, a date, and anything in between are all acts of improvisation.

CONFLICT AND NEGOTIATION

The skills and lessons from improvisation apply to many aspects of our professional lives, especially mediation and negotiation. Jonathan Bartsch is the President of CDR Associates, one of the nation's premier mediation and conflict resolution firms. He's had some interesting experiences with conflict resolution in difficult spots around the world: Pakistan, Sri Lanka, and Afghanistan, among others. We like to go ski touring in the winter and, after a blissful sunny ski day with some fresh powder, we started talking about how the role of conflict resolution and mediation is all about the art of improvisation — with very real stakes and outcomes.

"Facilitation brings together a unique combination of mostly powerful people who are used to getting their way," Jonathan said. "They get in a situation and they can't get their minds around listening or understanding other people. They don't know how to cooperate." He said part of his role as a mediator is to manage the inevitable clash that erupts. What is interesting, he noted, is that there is this attitude that can develop, *I'm dependent upon you to accomplish what I need to accomplish.* This dependency, which is mutual, creates space for cooperation.

I said, *When you sit down in a negotiation it can't be easy to ask, 'What do you want?'*

"That's not where you start," he said. "Nobody talks to someone they don't trust when they are in a conflict." Bartsch explained that

> People don't always know what their interests are. Through the Socratic dialogue process, you can better understand, and be understood by, others. You can also prioritize what you want. By listening and hearing. With people in real conflict, only when they've felt they are heard are they open to different ways of thinking about a situation. Without that, you can have the perfect answer, but it won't matter.

Jonathan explained how listening and being heard both open the space for cooperation. As we were coming down the mountain, he added,

> Good negotiators, and good leaders for that matter, go beyond listening, and, at the right time, chart a path forward that identifies the methods and/ or substantive paths that will meet as many of the interests as possible in order to advance toward resolution.

In our conversation, Jonathan echoed lessons from Alissa Ahlberg about the act of improvisation. Jonathan talked about how successful negotiations and mediations of big public disputes depend upon active listening. *Listening to understand versus listening to respond.*

Jonathan hit the bullseye. In their all-important face-to-face interactions, good leaders probably rely on one skill more than any other: *Active Listening.* As we watch someone closely, we start not only observing their comportment, but also instinctively taking it on, unconsciously mimicking their tone, gestures, and movements — even their tics. This allows our brain to develop neural resonance, which, as previously noted, is closely tied to emotional intelligence. Again, we find ourselves back at empathy. Some may dismiss it etheral, but empathy can be thought of as a brain function that improves socialization. You can conceive of it as a muscle to be exercised and strengthened.

Mediation is great, but when do you take a stand?

This happens not only in mediations, but also in hostage negotiations, which are the same thing, just with higher stakes and more pressure. Chris Voss, our FBI hostage negotiator, writes, "Negotiation serves two distinct, vital life functions — information gathering and behavior influencing — and includes almost any interaction where each party wants something from the other side." He explains that nearly everything in your life — career, relationships — depends upon this skill. Negotiation, he says, is "nothing more than communication with results."

We are constantly making choices. To prime ourselves as decision makers, we can (and should) apply the OODA Loop and the Art of Improvisation to communications strategy. In mediation, a negotiation, in a courtroom, on stage, in the classroom. We can slow it down for a more methodical approach for long-term planning. Or we can speed it up minute-by-minute. Or, if you're someone like basketball ball legend Michael Jordan, even millisecond-by-millisecond in rapidly evolving situations. The point being, the OODA Loop gives us a framework for responding to complex situations and allows us to iterate and adapt to the unexpected. Approaching challenges and conflict with a creative *Yes, and...* mindset enhances our ability to improve and resolve matters. Remember, it is HOW we make our choices that defines us.

THE ART OF THE MEETING

1. Take three deep belly breaths
2. Orient
3. Interest: check who is there and why
4. Take measured breaths
5. Observe how people are reacting
6. Decide: does the speaker know what they are talking about
7. Offer something of value
8. Make another assessment
9. Repeat often

FIGURE 9.3
Meeting meditation.

BREAKOUT SECTION: THE ART OF THE MEETING

The Art of the Meeting

Mark Williams and I have talked about applying Jedi-tricks to a meeting. The "meeting" is a great example of using improvisation, decision making, facilitation, and negotiation in an everyday real-time situation. Call me crazy, but I love meetings. Wait, I'll qualify that: I love meetings that *serve a purpose.* Meetings are one of my very favorite tactics and ways of connecting with people. Meetings are usually the epicenter of decision making. As fellow communications strategist Curtis Robinson likes to say, "meetings *ARE* the work." The best way to build a coalition, mediate a conflict, or engage in negotiation is to put everyone in a room together (or, at least, in the virtual meeting). Meetings are where matters are explained, decided, and delegated.

Next time you go into a meeting with multiple participants, try this exercise.

- As the meeting begins, take **three deep belly breaths** to calm your parasympathetic nervous system. Orient yourself.
- When the meeting begins, unless you have to speak first, focus that flashlight of your mind onto the person who is speaking. Just **listen and observe**. Don't even write anything down unless it's mission-critical. Stay hyper-focused. Don't pick up your phone or tap away on your laptop. This is **Active Listening.**
- Now briefly **scan** the other people in the room. What are their interests? What brings them there? What do they represent? Assess their role. Avoid judging them and imposing your own biases on their positions. People often form opinions without giving the other person much of a chance.

- Continue to take measured breaths.
- **Observe** how people are reacting. Are their hands on their face (a sign of boredom)? Are they looking at their phones? Daydreaming? Reading something else? Or are they as hyper-focused on the speaker and the meeting as you are?
- Bring your attention back to the speaker. **Decide:** Does the person know what they are talking about? What is their contribution? We all know people in meetings who sit there stone-faced. It may look like they are paying attention, but they're probably thinking about all the other work they should be doing.
- What about when that speaker is **US**? What unique gifts and contributions do we bring? What do people expect of us? Check your own body language and posture for the nonverbal signals you are sending. When it is time to **Act, offer something of value**, say your piece succinctly and with confidence (even if you do not feel it). Then be quiet and listen.
- After speaking, start another loop with **another assessment**: How are people reacting to what you just said? What do their nonverbals indicate? What is the energy in the room?

The importance of this exercise is creating awareness of other interests while expanding our own view. We become more mindful of other participants' reactions and the rationale behind them. Remember, the reaction defines the event. Through this "meeting meditation" we gain a keener understanding of our own interests and positions.

There are a million reasons to tune out in meetings, especially virtual meetings. Your mindset should be alert to the slightest change in the elements and sensory information. While you're still and quiet, stay attentive. Be ready to toss in your ideas and thoughts when the timing is right. If the CEO wants to know what you think about a specific issue, it may very well be when you least expect it.

10

Working the Room

The human understanding when it has once adopted an opinion… draws all things else to support and agree with it. And though there be a greater number and weight of instances to be found on the other side, yet these it either negates or despises… in order that by this great and pernicious predetermination the authority of its former conclusions may remain inviolate.

Francis Bacon, Novum Organum, 1620

Commander's Palace isn't just any restaurant. It's perhaps the most fabled fine dining restaurant in New Orleans, a city known worldwide for its creole cuisine. Cradled in a 180-year-old oak tree, the Garden Room is just one of its elegant dining rooms. The room, all glass, overlooks a classic New Orleans patio with wrought iron, ferns, and a flowing fountain shaded by live oaks. Lafayette Cemetery across the street was the legendary home of the vampire, Lestat, from *The Vampire Chronicles* by Anne Rice. The mansion and gardens are steeped in New Orleans history. In the early 1990s, the corner table in the Garden Room was the most requested table in the country. And I was the one answering the phone.

Ella Brennan was the matriarch of the restaurant. She won Restaurateur of the Year and the James Beard Award for best restaurant of the year in 1996 and the Lifetime Achievement Award in 2009. One of the great ladies of our time and a powerhouse, Hurricane Ella, as she was sometimes known, took me under her wing.

My first job after college was as the assistant *maître d'*. I spent most of my time booking tables and filling the restaurant to capacity — but not *over* capacity. Emeril Lagasse, the chef at the time, had a temper that was as sizzling as his cooking. If the kitchen was ever over-stressed because we sat too many tables at once, we felt his wrath like a two-ton meat tenderizer. He later became one of the most famous celebrity chefs of all time, with his own cooking show, books, spices, and successful restaurants around the country. In the kitchen at Commander's he would develop his signature catch-phrase, "BAM!" (which, incidentally, conveyed a lot of meanings). For instance, Bam! could mean he was particularly pleased about adding a bit of spice, whisking up a certain creaminess, or achieving a perfectly crispy texture. It could mean Bam! I'm going to knock you upside the head if you keep overcooking this dish! Or late night among the kitchen crew, it could mean something a little more fun.

"BAM!"

In the '80s and '90s, it was the exclamation used to punctuate the kitchen clatter at Commander's Palace like a firecracker. BAM! was about to make this grand old dame the number one restaurant in America.

One night, Ms. Ella said to me, "Son, I'm going to teach you how to work a room." She walked me around and showed me how to strike the perfect balance when approaching a table. You needed to be friendly to make guests feel special, but with a light touch so that you didn't intrude on their experience. An improvisational waltz. An evolving OODA Loop of entertainment. Guests want to enjoy their meal while also feeling welcomed — and maybe a little special. They want to feel that you have everything under control and are looking after them with a keen eye. These warm tableside exchanges require a touch of quiet confidence and, more than anything else, a heightened sense of Situational Awareness.

Ella taught me that I wasn't just an assistant *maître d'*, I was part of a team orchestrating a complete fine dining experience — one that went well beyond serving a nice plate of food. It didn't matter who the patron was, we treated them all like stars. And often they really were stars. Our guests included politicians, celebrities, and musicians. One night it would be Billy Joel, the next Adnan Khashoggi, the world's largest arms dealer. We extended to them the same warmth and congeniality we showered on the Old Line New Orleanians or a party of tourists from Oshkosh. Near the end of an outstanding documentary about Ms. Ella called *Commanding the Table*, she reflects on her storied career. She explains that Commander's Palace isn't just a restaurant, it's a metaphor for how you should treat people in life. With respect and graciousness. As if you were welcoming a stranger into your home.

I've come to believe that everyone should have the experience of waiting tables or working in a restaurant. It teaches organization and humility, introduces you to wonderful people (and, sure, a few unhappy ones), gives you a few headaches, but most of all, it shows you that there is no higher purpose than service to others.

My experience at Commander's Palace offers a fundamental communications rubric and illustrates why we should think of communication in terms of a quantum structure. As *maitre d'*, I could be welcoming or off-putting all in the rise or fall of my tone. Like the classic scene in *Ferris Bueller's Day Off*, when Ferris impersonates Abe Froman, the Sausage King of Chicago, at the upscale French restaurant: "Snooty?" Ferris challenges. "Snotty!" hisses the obnoxious gate-keeper behind the podium.

I would have been fired if I had acted like the *maitre D'* in the movie. The point being, our perspective depends upon where we stand and what our interests are at a given moment (Relativity). Every exchange is always dynamic, never static. Approaching every table entailed a different OODA Loop. What interests us one moment may bore us the next. We are not dealing with fixed points in a static universe. Rather, we must account for a world in flux, one in which perspectives and relationships are shifting from one moment to the next. Under such slippery circumstances, we should be aware of how delicate both ends of the communication process — the sending and the receiving — really are. In this respect, my time at Commander's Palace also revealed how decision making of all kinds, including something as simple as how we conduct ourselves in the presence of others, is predicated by situational awareness. Think more specifically of those two O's in the Loop: Observation and Orientation. (These also have a lot to do with our Locus of Control.)

I have observed that acute situational awareness is a common trait among successful, actualized people. The Engaged Ones. To understand how to succeed, one must first understand the situation: the norms, rules, players, interests, resources. It's why some people may not know much about history, biology, or geometry, but they can still control The Turf. They may be Freemasons, Grand Masters, union bosses, presidents, and CEOs. When we talk about "street smarts," we're talking about situational Awareness. With a capital "A." Hunters and woodsmen, with their heightened sense of smell, timing, and movement, have very acute situational awareness of their natural environment. In short, when the stakes are high, when it's either kill or be killed, few attributes will serve you as well as Situational Awareness.

WE HEAR WHAT WE WANT TO HEAR

In the previous chapter, Mark Williams compared our Situational Awareness to a mirror that we need to constantly clean and polish. To do so, we must keep our biases and assumptions to a minimum. As I've said before, life has a frustrating tendency to disregard our plans; rarely does it pay any more respect to our expectations and prejudices, not to mention our hopes and wishes. At the same time, an optimistic hope-dies-last attitude can be an important virtue. So, how do we "look on the bright side" without succumbing to self-delusion? Admittedly, these are tough questions.

Simon and Garfunkel once sang that *a man hears what he wants to hear and disregards the rest.* This isn't just a phrase, it's a profound statement about our situational awareness. A concept backed by science that shows the human tendency to see only what we choose to see.

Amos Tversky and Daniel Kahneman went on to win a Nobel Prize for their work in applying insights from psychology to economic theory, particularly in decision making in times of uncertainty. They studied how the rational brain — or, more accurately, the irrational brain — can distort our perspective and influence our emotions. Their ingenious research showed how we have "cognitive biases," unconscious errors of reasoning that distort our views of the actual world. They demonstrated that people do not behave in the way that economic models traditionally have assumed. People do not "maximize utility." Tversky and Kehneman developed Prospect Theory, an alternate view of decision making based in psychology that identifies how these biases affect our choices.

In my own conception, I like to envision a lake divided into four quadrants. One is windy, one is sunny, one is rainy, and the last is cloudy. We may try our best to visualize a lake that is cloudy and rainy, but what we see, what our brain *wants* to see, is the part that is sunny. In general, our brain has a tendency toward optimism — and to such an extent that it borders on delusion. There could be a storm moving in, threatening wind and rain, but if there's a hint of sun on the horizon, the latter is what we'll focus on. Or maybe we do focus on the rainy part because it reminds us why we never wanted to come to the lake in the first place and reinforces our desire to go home. Science shows our brain collects information to support our perspective and ignores the rest, as true as it may be.

This bias applies to just about every aspect of our existence, from careers to relationships. We observe situations through certain prisms. The way we approach a scene colors it all. But can you ever remove these filters and see the world objectively? Probably not, but what matters is that you are *aware* of the filter you are looking through and accounting for that — like accounting for sight off-set on a rifle.

The tendency to see what we want to see is called "motivated perception." For instance, in this social media age when we're always reminded of what everyone else is doing and so many users glamorize their updates, many of us suffer from FOMO (Fear of Missing Out), which diminishes our fulfillment of whatever we're doing at the moment, however enjoyable it might be intrinsically. People tend to want to believe that another part of the lake is sunnier and more fun than where we are.

FIGURE 10.1
"Four Seasons' Lake."

Of course, (borderline) delusional optimism may not be so bad. If this kind of internal narrative helps us muster the fortitude to persevere in difficult situations, it may just be an asset. Perhaps, over millennia, it has even functioned as an evolutionary advantage. After all, optimism underlies the willingness to believe in a better future, to believe there is a reason to keep going. It is in our nature to endeavor, not only to survive in a hostile and chaotic world, but also to thrive in it. We must soldier on because to believe that the world is just wind and rain and no sun is to give up on life. Hope can die. *But that is not our way.*

Accordingly, it isn't the stories we tell others, but **the stories we tell ourselves** that are the most important of all. While they may cross over into delusion at times, these stories are some of our most vital survival mechanisms.

So, let's talk about our *internal story.* Situational Awareness is taking information, observing, and orienting to create a story. What do our inner conversations have to do with our social intelligence, how we socialize? As Dr. Cuddy discovered, just about everything.

The secret is not to try and eliminate these filters, because that's impossible, but to recognize the kind of filters we put on our internal conversations. Once we can identify the kinds of glasses we put on to look at the world — the color and clarity of the lenses — we can be better at interpreting it.

In working the room with Ella Brennan, I learned that our social intelligence is formed by inner conversations related to our reactions to little stimuli. (*Do they not seem to be enjoying their food? Are their drinks empty? Are they anxious or impatient?*) Our internal stories and those we tell others are like two sides of the same coin and work in tandem. They are dependent on each other.

The story we tell ourselves is governed by the charioteer — we are in a constant struggle for balance between our two horses of reason and emotion. This dance — an engagement with the unfamiliar faces, full of unscripted moments and underscored by the intention of making everyone feel welcomed and special — was the Do-Si-Do Ms. Ella taught me on those magical nights. We were there to add *Care* in the *City that Care Forgot.*

Identity is the story we tell ourselves that is projected on the outside world. Who we are also depends on the story we tell others. And the one that others tell about us. This externalization of inward values and priorities, to a large extent, forms our identity. Your emotional intelligence is determined by your awareness of your own identity. We've all met people who have a very strong sense of self, and surely enough, we've met just as many who struggle to define themselves. The point is that to work the room you must comprehend the emotions of those in the room.

In all of our endeavors and experiences, serving and being served, *We are always working the room.*

Part 2

Communications Planning and Implementation

11

Where Are You Going?

The first step in writing a Communications Plan is to address intention, goals, and objectives: *Where are you now?* and *Where do you want to go?* If you are at Point A, where is Point B? Sounds simple enough from the perspective of an individual, but it can be surprisingly difficult to give a definitive answer as an organization. Members of any group may have different interests and agendas. Even a defined shared goal can mean different things to different people. So, identifying what they *don't* want can be a vital part of the process. I've observed various instances in which an account manager on a large account thought all objectives had been achieved and deliverables checked off, only to find out that the client had expected something quite different all long. Nothing ever turns out exactly as planned, but it is necessary to all agree on an endgame. (But keep alert: through the process of execution, endgames may change as new courses are charted.)

Only by clearly articulating goals can the right message be composed, the appropriate audience defined, and the tactics that get us to Point B identified. If everyone agrees on what Point B is exactly, then you have a clear bullseye on success. Perhaps you're trying to change policy, increase visibility for a new CEO, manage a sensitive issue with a city, build an art project, or elect a candidate. Fundamentally, these are all exercises in communication, and communication begins with knowing what you want. And that also means knowing what you don't want. All of this is to say that the most efficient way between points A and B will only be found by planning.

Marshall Ganz proposes that "to answer the 'why question' — why this matters, why we care, why we value one goal over another — we turn to narrative. The why question is not why we think we *ought* to act, but rather, why we *do* act, that which actually moves us to act — our motivation and values." Simon Sinek, who wrote the insightful and aptly titled book, *Start with Why*, also explores how good leaders focus on this simple, timeless question. He concludes that the answer drives everything else.

In the World Health Organization's *Effective Communication Handbook*, it is stated that in developing a Single Overarching Communications Objective (SOCO) for any type of communication, the most important questions to ask are these:

1. *Why am I speaking, writing, answering, or presenting?*
2. *What is the change I want to see as a result of my communication?*

Before setting in motion any game plan, you must examine the intentions. When making a wish, ask yourself: why *this* wish? What are your motives? Some may be tangible — sell more products, make a living, or protect a friendship — but sometimes, upon further examination, the real motivation may be ego and insecurity. *What real purpose will our goals serve? Why do I (or the organization) care?* Are the goals born of desire, jealousy, longing? Or do they stem from the more fundamental realities at the base of Maslow's hierarchy of needs, such as shelter, hunger, and thirst? Or is the intention to add something of value, however minuscule, to the common good?

There is a Sanskrit word, *Sankalpa*, that roughly translates to a vow or intention to move forward with a focus on the *nobility of the effort*. One must get one's self out of the way and concentrate on this larger Call to Action. Then it is possible to push away the inevitable self-doubt, the fear that you won't be able to properly articulate what you have to say, or worse, that you have nothing to say.

Every communications plan needs to be thought of as a campaign in advertising, diplomacy, or policy. The objectives and goals have to be clear — even if only to yourself at first. Otherwise, go no further. Any communications plan that tries to move forward without direction is doomed to stumble and fail in the end.

Only after you know exactly what the end is can you determine what the right means are. Then a message — a particular story tailored to a particular audience — can be created. Such tailoring requires selecting the right tactics, striking a balance, and finding the best interplay of resources. This is where the "art" of your communications plan comes in. With each new project, you are bringing into the world something that didn't exist before.

Once we clarify our intention, we need to create a Communications Plan. For every client and candidate, from international corporations and nonprofits to the White House, it's imperative to always ask yourself these three simple questions:

"**What** are we trying to say?" (The message)
"**Who** are we saying it to?" (The target)
"**How** can we say it?" (The tactics)

12

Reason and Emotion in Stories

WHAT WILL YOU SAY?

All the shampoo in the grocery store aisle are essentially the same, perhaps with a few differences in color and scent. Whether the product is eco-friendly, the cheaper off-brand, or "specially formulated" for extra luster or dandruff, at the core, it's all the same: soap. Maybe you pick one because you feel you deserve a shinier coif. Or maybe, as a college kid on a budget, you're willing to pick the cheaper off-brand. Maybe a pink bottle appeals to you. The reason you purchase one over another is that it tells you a more appealing story that connects with what matters to you.

The brain is not a computer. It is not mathematical, calculating and logical without exception. What separates our brain from those of other species is our capacity to tell ourselves stories that give us emotional cues for how to feel about something.

While getting my degree in economics and master's in public policy, I noticed many professors tended to believe in rational actors. For decades, economists and social scientists assumed that human behavior was predicated on reason. Economists had developed and adhered to a theory of *rational choice*, in which a person weighed all opportunities and challenges and made a specific calculation of the utility of an option before making a decision. After dutifully compiling all available factors and context and processing it all, economists assumed these actors spat out the *right* decision. But doesn't such an assumption rest on the notion that our brain *is* a computer, albeit one made of flesh, composed of neurons instead of microchips? This calculation assumed we were, more or less, rational actors in the game of life, making economic decisions based on a strict cost-benefit analysis. The problem is, when you get out in the real world, *rational choice* is not how we actually operate.

These rational choice "experts" were overlooking emotion.

In *The Political Brain: The Role of Emotion in Deciding the Fate of the Nation*, Dr. Drew Westen concludes that voting has everything to do with emotions and very little to do with rational choice. Outside of theory, I'd say real life has shown that voting and reason barely even know each other. I've seen countless examples of people voting or acting irrationally, hitching their fortunes to a particular narrative that appeals to their emotions, if nothing else. They go for the story that best reflects their own self-image — the narrative that speaks to the ideals and values they most closely identify with — even if those ideals have little to do with their daily reality, even if their favorite candidate's policy positions go against their own self-interest.

These actions reveal the complicated psychology of the voter and consumer. All too often, they aspire to be something or someone other than who they are. They support policies that are not in their immediate best interests, but in the interests of the avatar they see themselves as. Put another way, *they might be earning minimum wage, but they'd rather think like a millionaire.*

Remember the idea about our *awareness*, or consciousness, as the chariot driver of two horses, reason and emotion? Let one get away from the other and the chariot gets out of whack fast, sometimes with devastating consequences. Think of suicide bombers strapping explosives to their chests. In no way are they blowing themselves up because of something rational. Rather, they are acting out of strong emotion tied to ardent belief. They have been convinced of a higher purpose. They blow themselves up because they believe in a better story. They don't believe the Dalai Lama's assertion that the purpose of life is happiness. It's the same with early Christian Scientists who refused medical treatment on religious grounds. It is why domestic terrorists attack the US capital.

The irrationality of humans is a well-known fact and scientifically proven. Basic psychology provides three ready examples: (1) Cognitive Dissonance, which is the holding of contradictory beliefs and values, (2) Confirmation Bias, the tendency to seek out and uncritically accept sources of information that align with our established views (just think of social media), and (3) the Illusory Truth Effect, which essentially shows people will often go with the familiar choice rather than the demonstrably better choice.

But the superior narrative wins more pedestrian battles as well. Talk about education funding and the budget stabilization factor or transportation budget shortfalls and watch people yawn. Tell them a good story about kids in classrooms without teachers or teachers sleeping in their cars because they can't afford rent, and they're much more likely to pay attention. Show them bridges collapsing and suddenly their eyes light up. Dr. Westen found that when you talk about the technical aspects of an issue certain parts of the prefrontal cortex light up. But when you tell a story, the entire brain is stimulated. We are hardwired to understand stories better than we do facts because we personally relate only to the former. Westen refers to stories as competing alternately in the Marketplace of Ideas and the Marketplace of Emotions.

As Westen noted, our brain is a giant storytelling machine. Our ability to win the hearts and minds of family, friends, colleagues, and clients is based on our ability to tell good stories because our brains are hardwired to understand and respond to stories. Our brains take bits of information — data, facts, and stimuli from the external world, such as weather — mixes that with our mental record of everything that has ever happened to us and tries to make a narrative of it all. There is also the power of culture that shapes our beliefs about what is "good" and what is "bad." This creates unconscious heuristic shortcuts that profoundly impact our decision making, many times without us even realizing it. Our consciousness is telling us a story that defines our experience. And this story is always, always changing. Forever in a loop.

Messaging is the term of art used by communications professionals to describe a particular story, namely a set of thoughts surrounding an issue or an experience that a particular entity — be it a company, candidate, or advocacy group — wants to promote. Whether it's writing a play, a song, a speech, an advertisement, or a resume, the best ones — the ones that move us to do something — tell the best stories. It may be a story about why the target audience needs a certain product to solve one of life's little problems, like greasy hair or razor stubble. The more a story speaks to us and fits the stories of our lives, the more convincing it is — and the more likely we are to make the purchase. Messaging, if done well, can even ignite passion and lead us to commit to something greater than ourselves.

The assignments and tasks we take on as communications strategists are wide and varied. You might be required to create messaging for a ballot measure for broadband or for water usage or a community issue regarding transportation. You could be called on to triage emergency communications for relocating 80 senior citizens displaced by a fire who are currently standing in their pajamas outside with no place to go. Or maybe it's developing talking points for an obscure policy issue on school lunches that seeks to remedy the shaming of kids who don't have lunch money. Perhaps a head researcher stops by your office, drops six binders of an Environmental Assessment for a new transportation project, plans, and public comments on your desk, and asks you to condense it into four pages for public consumption. Anything is possible.

I remember a particularly tough assignment from when I was working with the White House as the deputy press secretary for the Denver Summit of the Eight. Nearly every one of the world leaders who gathered for the confab smoked. Boris Yeltsin from Russia, Ryutaro Hashimoto from Japan, Jean Cretien from France — they all smoked cigarettes (Bill Clinton preferred cigars). During the meeting at the Denver Public Library, they could be seen puffing away on a little balcony off their meeting room. As reporters pointed out, however, the policy at the library was No Smoking. I was given the unenviable task of writing the special-exemption smoking policy for the Group of Eight. The statement was something akin to the pigs in George Orwell's Animal Farm, "All pigs are created equal, but some are more equal than others." I received good-natured ribbing from my friends in the press corps for the mushy explanation. As I said, often our role is to absorb friction. Sometimes you're going to have to be the fall guy. Develop a thick skin.

In order to come up with a message that fits the situation and leads to the optimal outcome, you need to put a few things in place. First, you must define the issue, which usually entails research. What has happened in the past and how have people reacted? Which interest groups will be impacted? How will friends and family react? You must also understand the organization's *values and objectives* (not always an easy task, especially if the organization is unclear about its own purpose), know what worked and what didn't about past messaging. For messaging to work, the audience must first receive it (repetition helps), then understand it, and lastly accept its relevance (that means not just agreeing with it, but believing in it). Only then will it be possible for an audience to take action.

Before producing a Communications Plan, consider using a tool for specialists called a SWOT analysis that assesses *Strengths*, *Weaknesses*, *Opportunities*, and *Threats*. The SWOT analysis is a framework for both widening and clarifying your view of an issue, for becoming more aware of all possibilities. A SWOT analysis is a good way to assess where you are in relation to where you want to be and how to get there. Whole books have been written about just this process but the basics are simple. To advise a client, you must first get the lay of the land. Through this process, you come to a greater understanding of the historical record, context, skill of the current team, personnel issues, and potential areas for crisis. This analysis will help you determine your perspective and your *approach* to the issue. This kind of analysis is the first step in creating a roadmap.

The process of research and discovery may also involve polling or some form of opinion gathering. Research, polling, focus groups — they all have one overriding purpose: to divine public opinion. Presumably, if we know what people think and which values they hold dear, then we can create a pathway to caring. If they care, then they will purchase the product or vote for the tax increase. Measuring public opinion may also offer a baseline for where you are. Point A. You can use polling to test potential messages. You

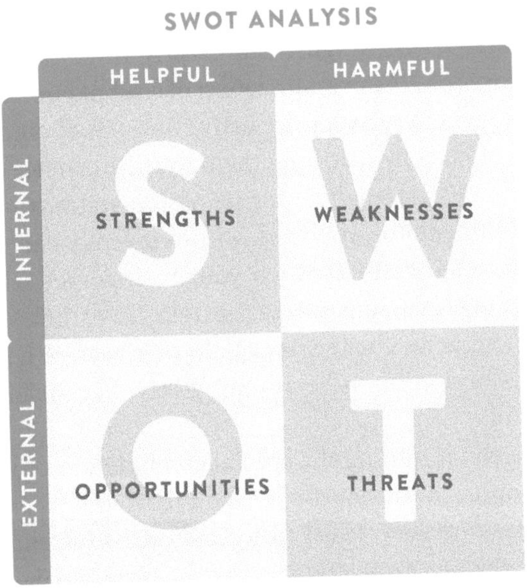

FIGURE 12.1
SWOT analysis.

can also use it throughout a campaign to track perception to determine if, and how, your plan is working.

After deciding what objectives you want to accomplish, start your communications plan by focusing on the approach. What's the angle? Journalists and news editors would call this your *Point of View*. Are you going to respond? Not respond? Will someone else respond on your behalf? Will the response be aggressive or respectful? Defiant or contrite? Will you apologize for a perceived gaffe or deny it altogether? You can come at any issue from the front, the side, the back, or all around. The approach sets the *tone*.

Your outlook, or *perspective*, on an issue will determine how you approach it. How do you view the situation at hand? Are you feeling threatened, confused, protective, secure, hopeful? Rotate the map to freshen your perspective. Artists invert their canvases because it's easier that way to spot mistakes in perspective. Another approach is to base your game plan on your opponent's. Football coaches spend a lot of time watching tapes of opposing teams to learn their strengths and weaknesses. Then they devise their game strategies accordingly. Assess your own opportunities and threats as if you were watching them play on a screen. That's always an option for how to approach a situation.

After deciding on the goals and approach, the next part of any communications strategy is to figure out exactly what you are saying. *The Message*. Messaging is the crystallization of a set of circumstances into a coherent set of facts to relay a perspective on a situation. It makes sense of a multitude of facts by distilling them to their essence. I occasionally provide on-air political analysis for television stations during election season. As a commentator, you work within severely narrow constraints of about seven to ten seconds when your turn comes. There is usually only time for a declarative statement or two. To ramble on is to not be asked back. There is no chance to explain or waver. When commentating, on politics or sports, to be anything less than instantly clear and decisive is to fail.

Messaging is about finding the special combination of words to distill something in order to move people. Good messaging is a call to action. What if the message were tweaked to say something slightly different? Perhaps to emphasize a specific facet of an issue. Or perhaps to avoid or, at least, mitigate a perceived offense and still get the point across.

These days talking about *real estate development* may not gain as much traction as championing *affordable housing* because of the controversy surrounding gentrification. (This is the kind of sensitive area that can be identified if you have the resources for public opinion research, such as focus groups and polling.)

Most of the time it isn't just *what* we say, but *how* we say it. Actually, it may be more accurate to say that the *what* and the *how* inevitably bleed into one another. This may be another way of expressing my first belief: *The energy behind an act of communication is more important than the communication itself.* As Drew Westen and his fellow neuroscientists concluded, in any message, the words — and even the facts — are only as important as how they make people feel.

A key step to effective messaging is creating a messaging roadmap, including what kind of effect you want to have on an audience — and which methods and what content are even capable of producing such an effect. Many think that the purpose of communication is to inform and educate. While this is certainly a reason, it is not the ultimate intention. (In itself, education suggests a *reasoned* approach to communications rather than an *emotional* approach.) As expert speech coach Dr. Andi O'Conor maintains, the real purpose of communicating is to create a *relationship*.

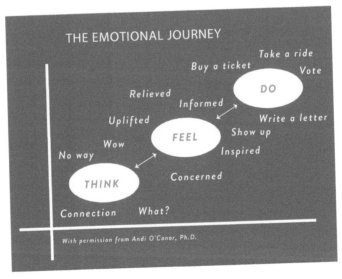

FIGURE 12.2
The emotional journey.

In crafting any newsletter, email, speech, website, or video, Dr. O'Conor advises to think first about the audience, then create an intention. This intention should ask:

What do you want the audience to
Think?
Feel?
Do?

From there she recommends creating an Emotional Map. A journey of where you want to take the listener.

13

Messaging and Movements

In order to increase the likelihood that a message is effective, one can rely on a series of storytelling exercises devised by our master organizer and expert on public policy, Marshall Ganz. He, too, was interested in the question of how to get people to care. More specifically, he was curious about what creates a social movement, who are the actors involved, what are their issues and passions, and how do they make change happen?

He contends that a *Theory of Change* starts when one asks: *How do things happen*? To make something a reality, to bring it from an idea on the page to fruition in the real world it has to become *tangible*. One can precipitate this change by crafting three stories:

The Story of Self
The Story of Us
The Story of Now

The foundation of any movement comes from identifying who you are and what you believe, the *Story of Self*. This is where we define our identity. The values that shape our choices, which define who we are. The experiences that make up our personal narrative. According to Ganz,

> What is utterly unique about each of us is not a combination of the categories that include us (race, gender, class, profession, marital status), but rather, our journey, our way through life, our personal text from which each of us can teach.

These categories, our demographics and background, form a personal context or backstory, but it's our thoughts, beliefs, decisions, and actions that shape and define our narrative most indelibly.

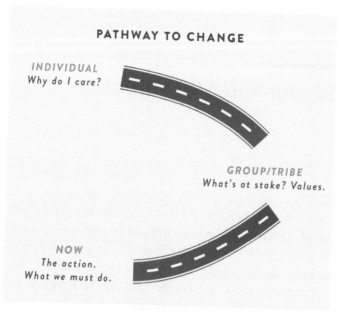

PATHWAY TO CHANGE

INDIVIDUAL
Why do I care?

GROUP/TRIBE
What's at stake? Values.

NOW
The action.
What we must do.

FIGURE 13.1
Pathway to Change.

Next is to define who *we* are. The *Story of Us* is about our collective identity — not personal identity, but the two are closely linked. We are not one-dimensional. There are many versions of "us." Facets, or little islands, of our collective identity that derive from our family, school, church, sports, friends, country, ethnicity, profession, literature, music, and anything else we share with others. Each of these facets is a repository of stories, challenges, traditions, transgressions, and triumphs. We tell these shared narratives again and again especially through celebrations (Fourth of July, Christmas, Thanksgiving, birthdays, weddings, and such). I have a particular affinity for these occasions, especially festivals, where people come to celebrate, share, break bread, and dance. To create an "us" out of a disparate group of people, there must be a storyteller or something like music (think of any anthem or folktale) that defines and interprets a shared cause or experience.

Lastly, there is what we, as a movement, *must* do. The *Story of Now* gives voice to our most urgent challenges. What values are at stake if there is no action? What are the choices? The risks? The challenges and opportunities? The spark behind *Now* is first believing that you have the power to make change in the first place and then deciding that you must use that power. *Are you ready to take the journey? Now* usually involves a *prodromal* choice. *Are we going to put up with declining education funding forever — or is the time to act now? We just can't take it anymore.* The Rubicon has been crossed. The story of *Now* is the vision of how we move things from where they are to where we believe they should be. This is where we define the social change we want to see.

Some might argue that our personal stories don't matter all that much. Ganz contends, "If we do public work, we have a responsibility to give a public account of ourselves — where we came from, why we do what we do, and where we think we are going." These stories of *Self, Us,* and *Now* create a context for others about why we care so fervently about an issue.

Long before Ganz, Aristotle provided sage insight into communications, which any and all communications strategists ought to consider when putting together messaging. Aristotle taught that rhetoric consists of three elements: *logos* (logic), *pathos* (feeling), and *ethos* (credibility and legitimacy). Furthermore, he declared that stories (and by extension, messaging) consist of a Plot, Character, and Setting (time and place). The Plot, the series of events that constitute a story, is essentially the framework that establishes what's at stake and why it matters. Characters then fill in the plot with their unique interests and points of view. As Aristotle wrote, "The protagonist's tragic experience touches us and, perhaps, opens our eyes." The setting lends the story a foundation and context by coloring in the world around the characters. All great stories have a unique and memorable setting. *Star Trek*, the Bible, the great novel about New Orleans, *Confederacy of Dunces*. The setting brings characters to life and gives them a world to live in, the characters give the story voice, and the plot provides meaning.

While it might sound like the description of the sweeping scope of great novels and movies exclusively, the structure also applies to advertisements, tag lines, brands, products, and political speeches. Support or oppose something, vote for this or that, show up at a town meeting about a neighborhood traffic circle. Theory of Change can give voice and meaning to create a movement about anything you can think of: sexual harassment, police brutality, climate change, immigration reform, health care, you name it.

Effective messaging must be *relevant* to ignite caring. Why is the issue of public education important to people's lives? Why should people vote? And for whom? What is important and why? The issue at hand must have social significance; it must be greater than any one person. A message does not exist in a vacuum. It isn't independent of the intended audience.

An effective message must be *memorable*. It needs a certain stickiness to stay with people and encourage them to take action days, weeks, months, even years after they first received it. In this regard, market or voter research can be particularly helpful in testing messages. Messaging goes on to form the basis of a whole campaign and is used in advertisements, social media posts, literature, attracting attention everywhere it can take life. Often the "messaging" document that gets approved will evolve into a press release, fact sheet, advertisement, slogan, or some other kind of public document. It can be translated into content for a website or a social media page. (At this point, it would make sense to double-check that the whole team understands the core message and is singing from the same sheet of music.)

Whether we're kicking off an important meeting, taking an incoming press call, doing a TV interview, or preparing a speech, we always do a quick 10–15-minute prep session with clients in which we focus our intentions by literally drawing a triangle. The Message Triangle is a helpful way to organize thoughts and craft a response that is tight, clear, and purposeful. This is not just another training tool for novices. We insist upon it for even the most seasoned CEO. The heads of organizations I work with realize the value of getting it absolutely right when they are on the spot. When the cameras are rolling, there is only one chance to make it count. Our charge is to deliver the very best response.

The points of the triangle are formed by answering *Who, Why,* and *How.* But also notice how nicely these overlay with the Theory of Change and the stories of *Self, Us,* and *Now.*

The Message Triangle also lends a small measure of control to the interviewee. Many think the interviewer asking the questions is the one in control. They're wrong. The Message Triangle allows us to pivot onto our own turf by a simple twist: *That's a great question about X, but what I'm really here to talk about today is Y.* Then we start hitting the three points of our triangle. Also, the beauty of the Message Triangle is that we can start at any point we feel is most effective.

This helps people who tend to talk *around* an issue and babble at the edges. They think listeners will connect the dots about what is important. But it is our duty as the speaker to connect those dots for them. Get to The Point. What is the Single Overarching Communications (SOCO)? When speaking to the press or answering questions in a forum or creating

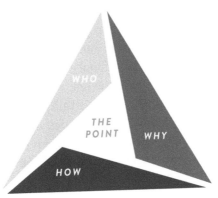

FIGURE 13.2
Message triangle.

talking points or presenting an advertisement to clients, we only have one shot to get it right. Make it count.

Here is how it can be applied to an association of teachers and educators:

Who: We are an association of 39,000 educators and the leading expert in the state on education matters.

Why: We are committed to ensuring every student thrives by providing the very best learning experience.

How: We use our collective voice to support better classrooms and influence education policy on issues such as school funding, teacher shortage, support for educators, equity and justice issues, and accountability for charter schools.

"BE THE VESSEL"

Creating messaging, writing a story, or crafting a press release is never as easy as it sounds. It can be tough, even excruciating. A process of discovery and decision. A process of illumination and elimination. Often you don't know what you're going to do until you're forced to write it down and put it out there.

Make no mistake, a communications strategist needs to be an artist one moment, a military general the next, and perhaps the world's greatest host the moment after that, improvising according to circumstance and situation. In one sense, you're a scientist methodically devising and surgically implementing a plan. In another, you're an artist trying to realize the vision with every word on the page and stroke on the canvas. It doesn't mean you have to know everything about everything. This is a key distinction. You're not the expert; you're the one compiling expert thought. As the communications strategist, you have to have a 360° view, situational awareness of the available facts, a list of experts to call, and the ability to boil down jargon into language everyone understands.

Jennifer Wozniak is a senior vice president of communications for Xcel Energy, the number one provider of renewable energy in the country and a utility that spans seven states. It's a tough job and as demanding as it is dynamic. One of her primary responsibilities when writing messaging is playing peacemaker, and managing competing interests, such as legal, regulatory, government affairs, and local interests. For the most part, she is in a constant battle to inspire action among people who would rather do nothing. This is not necessarily their fault. It's merely a symptom of corporate culture. In just about every big business you'll find smart people who prefer to say no more publicly than is absolutely necessary. Wozniak knows this is precisely the wrong approach.

Communications directors always face this kind of resistance. Despite it, her team still has to create and distribute content. Bill inserts, email alerts, branding, marketing of new renewable energy products, press calls, responses to regulatory questions, and utility commissions in all the states they serve. Providing energy to customers is Xcel's primary job, but it also has to *communicate* about how it delivers that energy.

Wozniak explains that when you actually sit down to write messaging or a statement or press release, it isn't just about what you want to say or how you will say it. Because the "what" isn't usually fully defined.

> There are viewpoints from regulatory, legal, corporate, operations — and everyone has a different perspective. Sometimes the act of writing it down for public consumption forces a sort of 'squaring up' with all the parties about what is actually going on and how the deal will go down,

she says. And that's what makes this particular job so difficult. The wrong message — the wrong word! — can explode into a crisis almost instantly after it's released. Jennifer's role underscores the importance of the ability to understand and work with different interests.

Her piece of advice on surviving in the rough-and-tumble world of corporate communications is to develop a thick skin, don't take anything personally, and pick your battles very strategically.

"Lawyers, accountants, and engineers rule our world," she says. The toughest part of her job is striking the right balance between hairsplitting the meaning out of a press release or document and having it tell a good story. Her job is to help humanize jargon and insider lingo. Make no mistake, Jennifer loves her job and is incredibly good at it. This is where the creative part comes in. She believes corporate America needs more creative people who have a sense of how to relate to others. "We tend to get in our own heads and overanalyze everything. We have a lot of high-level executive wordsmithing."

Social media is no cakewalk either. Wozniak identified one of the most challenging aspects of it, "People don't care about the right things. People watch cat videos. They don't support newspapers. Most people are just uninformed. As a society, we have significantly dumbed ourselves down with the advent of the internet and social media." She also talked about the need to stay on the balls of your feet. To always be proactive. Creating and executing plans. She noted, "It's not really about always going after the big kahuna. Our success is a lot of little victories. The slow drip. Always keep pushing and getting your message out." Sisyphus' Rock.

It's tough to fill a communications role if you have a big ego since it's your job to make sure that the focus is on others. You are there on behalf of the bigger cause or issue. We talked about how those attracted to the glare of the spotlight, soon burn up like a mosquito in a bug zapper. Advocates who make their work about themselves usually meet with failure. The strategist must think about being a diviner. A voice for someone else. If you are a spokesperson or speaking out on behalf of a client or interest, you have to remove your ego. What matters is your skill, not your personal agenda. Your approach should be based upon the data, situation, and context. You have to prioritize what is best for the organization. Your goal is to put that into words.

Ideally, the CEO should look at your statement and say, "You said this better than I could have said it myself." (But don't hold your breath.) The collaborative process in writing and creating messaging is a painstaking, frustrating, and beautiful process, which ultimately leads to the best results.

When writing out communications copy, whether for social media, newsletters, or websites, create a messaging document that delineates the tone as a reference. It should include what words to use and not to use, what we believe in and value. This isn't a personal voice. To be the voice of the organization is an awesome — and humbling — responsibility. It should distill all known factors — the players, the interests, and all perspectives — into something precious and pure. You have to be both left-brained and right-brained.

As Jennifer says, "Be the vessel."

Good messaging, combined with good organization and appeal, is how candidates pull off momentum-generating rallies, entrepreneurs make persuasive presentations to potential investors, and traffic engineers give convincing explanations about where to build a bridge. The most effective communicators pick the best fitting words — out of

FIGURE 13.3
Checkpoints of good messaging.

infinite combinations — to most clearly express their thoughts and emotions on a particular subject to a particular audience. They refine and sharpen the message through discovery and debate. They pay close attention to language, diction, and syntax. Doing all of this, they utilize the most powerful tools at their disposal to control their environment. Messaging — *What we are saying* — is our first step in creating stronger connections.

14

The Hero's Journey

Fortunately, there is a tried-and-true template for creating great messaging and telling memorable stories. A framework field-tested throughout millennia of culture, from Gilgamesh to Greek tragedies, Shakespeare to *The Lord of the Rings*. For us mortals, the template can be developed, experimented with, tweaked, and used over and over again. Even on a daily basis — in all aspects of our life — for any story we could ever hope to tell.

After talking to our astrophysicist friend Dr. Jeff Bennett about entropy, the universe, swimming, and much else in between, he turned the conversation to the dichotomous nature of our reality. He explained that the first world is the physical world we live in, the one that consists of land, sky, rivers, mountains, the earth rotating around the sun... the entire universe, you could say.

The second world is made up of the stuff of human activity: social interactions, politics, economics, and communications. This is the world we create. In a very short time (relative to the timeline of human development) we have concerned ourselves more and more with the second world, and it has precipitated a profound shift in human consciousness. The second world depends on and could not exist without the first world, of course, but we seem to have lost touch with that fact. We may live *on* the earth, but we live *in* our heads. We've forgotten about our physical existence. The second world derives entirely from our mental activity. It is made up of the stories we tell ourselves and each other.

The first world is the objective reality. The water, the cycles around the sun, physics, mathematics. The second world is our subjective reality, which is everything we make up: constitutions, money, political parties, language, and so on.

One way we understand these stories is through what Joseph Campbell, a literature professor who wrote highly influential works based on his lifelong study of comparative mythology, called the *Hero's Journey*. Campbell's interpretation of age-old storytelling methodologies has shaped more of culture over the past 60-some odd years than even those who know his work well may realize. In his seminal works, Campbell connected the dots of human experience for a general audience, not an exclusive readership of academic peers, which helps to explain their popularity. In 1949, he published *The Hero with a Thousand Faces*, in which he laid out patterns that recur in mythology throughout time across cultures the world over. He referred to the summation of these patterns as the "monomyth."

The monomyth is a basic (and widely applicable) story model in which a hero is called to go on an adventure, meets helpers, surmounts obstacles, solves a crisis or wins a contest, and usually comes home transformed. The cycle is also a powerful tool that can be applied to messaging of all colors.

There are many theories and "formulas" for effective storytelling. Whole classes are taught and books are written on this subject alone. The city of Los Angeles is built on manufacturing stories, plenty of which (the best, I'd say) rely on the archetypes that Campbell identifies. Dorothy waltzing merrily down the Yellow Brick Road in *The Wizard of Oz*. Hagrid whisking his young charge away from the Dursleys in *Harry Potter and the Sorcerer's Stone*. In fact, George Lucas credits Joseph Campbell with the creation of *Star Wars*. Think of Luke Skywalker as what Campbell called the "archetypal hero."

While Campbell called it a *journey*, I think of it more in terms of a *cycle*. With cycles within the cycle. In the classic Hero Cycle, the protagonist leaves home — sometimes by divine calling, sometimes by kidnapping — for an adventure. Our heroine (or hero) travels to the edge of her known world and beyond. She meets interesting characters that may help or hurt her along the way. After an arduous journey, our adventurer faces a great crisis. Through this test, she is often forced to cross a "point of no return." Ultimately, she triumphs (well, usually…), acquiring wisdom that will help her community back home. Almost invariably, there appears at the end of the cycle a redemption scene, in which the protagonist realizes a greater purpose by finally coming to understand the wisdom of a parental figure.

It's called Reconciliation with the Father-/Mother-Figure. At the conclusion of *The Wizard of Oz*, Dorothy realizes how much she loves Auntie Em. In the final space battle in *Star Wars*, Luke turns off the targeting computer and, obeying father-figure Obi Wan-Kenobi's call from the beyond to *Use the Force*, summons the inner focus to hit his target and destroy the Death Star. These are pivotal moments of character transformation. Completing the cycle, the protagonist eventually returns home with some sort of boon for the community and is forever changed by the journey.

Archetypal heroes, life-or-death quests into unknown worlds, mysterious guides and helpers, monsters who guard treasure — they all have deep roots. Sure, there are almost countless ways to put your own spin on it, but the arc remains classic. Aristotle is credited as the first to write about what nearly all successful stories depend on: the three-part story structure. (He likely borrowed it from someone else before him.) In Part 23 of *Poetics*, Aristotle proclaimed that the plot of a *Tragedy* "should have for its subject a single action, whole and complete, with a beginning, a middle, and an end. It will thus resemble a living organism in all its unity, and produce the pleasure proper to it." Since then, scholars have debated whether this identification of a story's beginning, middle, and end constitutes one of the most profound — or most obvious — statements in the history of literary criticism.

If we apply Aristotle's terms to the Hero's Cycle, we see that beginning, middle, and end neatly correspond to *Departure*, *Initiation*, and *Return*. I think the first and third terms speak for themselves. Initiation, to be clear, occurs when the protagonist, having left the familiar behind, confronts unprecedented trials and tribulations in a new and strange world. She must develop new skills, acquire new tools, recruit allies, and fend off the wicked. This is when the heroine decisively commits to the journey and is initiated into the extraordinary world.

To the uninitiated, it may be shocking to learn that Luke Skywalker, Dorothy, Bill and Ted, Buddha, Jesus, and countless others share a similar storyline. Yes, the particular details (context and plot) of each are what make them so compelling, but the similarities are too profound to ignore. As many have realized throughout history — and Joseph Campbell knew well — there is something about these kinds of stories that appeals to us on an instinctual level.

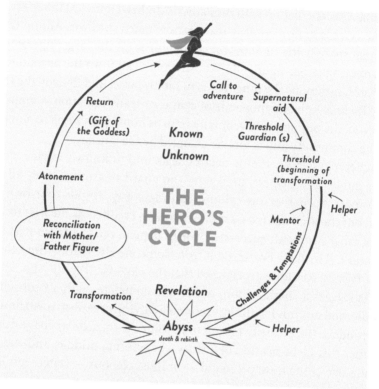

FIGURE 14.1
Hero's Cycle.

THE HERO CYCLE APPLIED

The Hero Cycle isn't just a way of telling a story, it's also a way of putting your own life in perspective. Here's how I might apply the template to my own daily life, specifically my swimming workout routine.

My own mini Hero Cycle begins at 4:40 am, when my alarm goes off. It's painful, but when I make it out of bed I think of it as having crossed the most difficult threshold. I rise and start making good Louisiana coffee. The morning papers arrive at about 5:00. I am out of the door at 5:20 with my gym bag slung over my shoulder. I step outside into the quiet pre-dawn dark. The workout begins at 5:40. Most hero cycles feature a strange land or mysterious universe. In this cycle, it is the enchanted world of water. My fellow swimmers run the gamut — former Olympians, college swimmers, seniors, and those just starting out. The warm-up begins and as the body starts moving, the water begins to feel better and better. Then the main set starts and its game on. My body and mind meet trials and tribulations. The intervals test my fitness and mental state. As the sets intensify, the dragons I face grow sharper fangs and claws. The question bubbles up, demanding an answer: *Will you push through to make the interval or give up and hang on the wall in defeat?* Inside me swirls a little fear, some self-doubt, and a strong sense of *get-your-ass-in-gear-and-let's-do-this!*

In the classic Hero Cycle, there is a reconciliation with Father/Mother-Figure. In a swimming workout, I've come to realize that reconciliation is within myself (is this what we face in all our pursuits?) Any worthwhile endeavor presents you with choices and decisions regarding what you *are doing*, what you *know you can do*, and what you *need to be doing*. These aren't exclusive. In fact, their intersection yields the best of all outcomes.

Some days I feel powerful and other days, I'm tired and sluggish. My body doesn't respond. No matter how I start out, I like to finish like a rock star. But it's also just a workout and we do what we can. Mostly, I try to just enjoy it and be grateful for being there at all.

Finishing the workout, I've been tested. I begin my journey home after a hot tub with teammates. Walking out of the gym, with the morning sun rising, I'm better for the experience. I've slain something. (Sometimes I'm the one who's been slain — and that's okay.) I'm stronger and wiser. Fresh and alive. A small Hero Cycle completed, and the day is just beginning.

While the Hero's Adventure is taught in film schools and literature departments the world over, the monomyth's beauty is in its nearly universal applicability — especially for the communications strategist. In applying the Hero Cycle to any situation, one of the central questions you should ask is, "Who is the Hero?" Many might assume that in marketing, political, and public policy campaigns, the product, candidate, or issue, would be the obvious choice. Sometimes such positioning works and sometimes it doesn't.

One rule of thumb is that when a major oil company becomes embroiled in some kind of environmental catastrophe, such as an oil spill, you can expect to see before long a media blitz of commercials and print and online ads vouching for the company's longstanding commitment to environmental stewardship. For example, after the 2010 *Deepwater Horizon* oil on the Gulf Coast, BP, which was ultimately responsible for the mishap, spent approximately $93 million on advertising from April to late July. The ad campaign drew criticism from public officials because a good portion of it was devoted to protecting the brand. TV spots consisted of pledges from the company to fund clean-up efforts and scientific research projects voiced over images of smiling fishermen and Gulf Coast wildlife. The commercials portrayed BP as the hero at the center of this tragedy, a noble character who may make mistakes now and then, but always proves goodhearted, community-minded, and nature-loving in the end. Literature investigating the effect of BP's "green" ad campaign has come to conflicting outcomes — some papers argue that it has redeemed brand reputation; others conclude it hasn't. At any rate, there may have been a more effective way of telling BP's story in light of the crisis.

In a marketing campaign, it's always good to consider whether the company should get out of the way and let the consumer be the hero, the one who chooses the right helpers to solve problems. Perhaps the heroine is the woman whose skin will shine with a certain luster when she uses a particular lotion. In this case, the advertised product, the lotion, is depicted as the helper. In the fight to increase education funding, we find something comparable. It is often more productive to portray as heroes, not the teachers who need increased salaries, but the students. Teachers who earn more are likely to stay in the profession longer, which means better student outcomes.

Similarly, you might assume that the candidate is the hero of a political campaign, but he or she is not. The hero is the *voter*. When people pull the lever for an issue or a candidate, they should view the candidate as the ship who will sail them safely across turbulent waters. As demonstrated in the application of the Hero Cycle to my swimming, you are only the hero when the story concerns yourself.

Again, this is a key distinction in the application of the Hero Cycle to Communications Planning: *Who are your Heroes? Who are your Helpers? What is the turning point? The reconciliation?* When creating messaging, consider the arc of the story and the distinct roles of each character.

The next chapter is an example *par excellence* of how to use the Hero Cycle to create an Emmy award-winning global advertising campaign with a simple butterfly.

15

Anatomy of a Butterfly

The reputation of a company or product is *everything*, and this reputation emanates from a company's story. The brand is the embodiment of all the stories that the consumer associates with a company — both good and bad. On the playing field of ideas and campaigns, if you fail to tell your own story, someone else will tell it for you, and I'm willing to bet you won't like it. A brand can be a "hero" within the narrative. Or, as depicted above, a product might be the "helper," who assists the hero-consumer. It's not just boring old soap. This particular soap washes away all the grime of life, leaving one feeling "fresh and clean as a whistle," as Irish Spring advertisements promise. It's the car that helps the Hero get the girl. The mortgage company that gives the Hero the keys to the American Dream of homeownership.

In the early 2000s, McCann Erickson, the advertising firm that made ads for Coca Cola and many other world-famous brands, spent months trying to come up with a new platform to promote Microsoft Network's *msn.com*, an aggregation website portal with an array of business and consumer services. The firm had recently taken Microsoft's old drab logo, which consisted of four blocks of colors — red, blue, yellow, and green — and transformed it into a butterfly, a major success. But it had run into difficulty with the platform for msn.com. Scores of ideas had come and gone. Some were way out there. Some were provocative and funny, but none were hitting the mark.

My old friend at McCann, Tom Giovagnoli, "Tommy G," had been in advertising for a long time. He is known for his zany, colorful humor, and tenacity on a campaign. He and his team had a new idea: to anthropomorphize the butterfly, to make it come alive with human qualities, and to connect people to the software. In effect, to *humanize* the technology. But with a nonhuman.

In its early days, the computer world was very cold and techy. Essentially, McCann Erickson was tasked with creating something warmer and more user-friendly. To make the winged insect more human, Tommy G worked with Jim Hensen's shop, which created the Muppets characters and brought to life some of the characters on *Sesame Street*. They outfitted actors in butterfly suits with antennae and wings in Microsoft's signature colors. Then they flew them into cities everywhere.

This campaign was successful and beloved because it was so simple, yet remarkable. The "Butterfly People" help you accomplish your online tasks so easily. When we dissect the anatomy of the campaign, however, we see the complexity: how the many communication principles we've discussed work in unison. Tommy G noted, "Ultimately, this is how good advertising works: a lot of information and ideas walking together through a simple and singular doorway." He pointed out that it never would have worked with a bunny or a smiley face logo.

The Butterfly concept evolved after several misfires. Before the Butterfly People, there was another campaign called MSN Project where a bunch of people lived together in a house with only MSN to get them everything they needed to survive. First problem, it was humorless. Then Tommy G did a global campaign of really charming stories of people using MSN in their lives. They were shot around the world and were well received, but people still couldn't wrap their heads around what MSN really was or essentially did because there were so many services.

With the Butterfly People, consumers finally understood that MSN was like your fun friend, a companion, and helper when doing anything online. The Butterfly People was one big idea that connected over and over again in different languages to different people with different needs.

Tommy G's Butterfly People ended up being one of Microsoft's biggest all-time consumer advertising campaigns. As Microsoft stated in a press release on October 14, 2002,

We've made a considerable investment in the marketing campaign around this launch; it will be our largest marketing spend ever, heralding the new theme, 'It's Better with the Butterfly.' When consumers see the MSN butterfly, we want them to know they're getting a superior service.

The global advertising strategy consisted of broadcast television spots, billboards, events, and the full suite of advertising channels. One of the brilliant grassroots tactics featured the new mascots rollerblading through major metropolitan streets saying hello to strangers, taking pictures with folks having lunch, flitting here, alighting there.

On a rainy afternoon brightened by margaritas and tacos, Tommy G recounted for me the final days before the pitch. With the team still debating concepts, Tommy G had an epiphany. He asked himself a very important question: *What is the simplest and most universal element of this MSN?* Basically, it's a bunch of internet services that organizes content for users. The single sign-in service was intended to facilitate the everyday operation of any business and provide related websites that addressed a wide range of consumer needs. When Tommy G thought of a worker in front of his computer toggling among documents, email, and websites, he envisioned the butterfly fluttering about from task to task. "The butterfly was the great connector. It evoked the freedom to explore and create literally on the fly," he said. The Microsoft Butterfly went on to become one of the more successful advertising campaigns in history.

Tommy G related the Butterfly to the Hero Cycle. Putting his own quirky twist on it, he described the idea as "the hero's mid-morning errand." In such a formulation, it wasn't the butterfly that was the hero. It was the consumer, who is unexpectedly thrust into a quest, perhaps one that lacks the dramatic flourish of rescuing a princess or saving the universe from a laser-beam-generating starship, but one that nonetheless rises to the top of our to-do list: making an online purchase or picking a movie for Saturday night. The butterfly serves as the helper/mentor and helps the hero through the early trials of the first threshold, which involves acquiring knowledge. As Tommy G explained, the hero needs such knowledge to establish some kind of order out of the "untamed wilderness of the digital world." By offering guidance, the butterfly helps the hero transform himself from subject to master.

While rain and margaritas continued to pour at our little taco joint, Tommy G talked about the importance of what our mythologist Joseph Campbell called an *Affect Image*, a visual that taps into emotions in a specific way. An affect image transcends language. For example, consider the way a simple sketch of a puppy dog pulls the heartstrings or what a peace symbol elicits or the blissfulness of a shot of a sailboat adrift on the vast ocean. Some of the strongest emotions can be evoked by images alone. And that is precisely how a brand logo works.

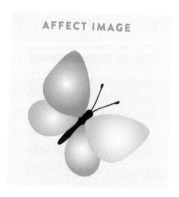

AFFECT IMAGE

FIGURE 15.1
Affect image.

Another way companies crystalize their messaging is through the use of taglines. An Energy company: "Responsible by Nature." A teacher's union: "The leading voice in Education – So every student thrives." I once created a tagline "The way to a better environment" for a composting service for a waste hauler.

I've come up with a simple rubric in the profession of writing: *The fewer the words, the greater their value.* Big, thoughtful books, sweeping histories, epic works of fiction get pennies for their words (if the authors are lucky…). Meanwhile, those in the field of branding make extraordinary amounts of money to come up with a brisk tagline and a simple little symbol. Delta Airlines' *We love to fly and it shows*, for example. Those seven words likely cost millions of dollars.

People who understand the value of the "experience" created around a product realize a better return on their investment. The better story they tell about a product, the more they can charge. Look no further than the outdoor clothing company, Patagonia. It makes exceptionally durable clothing for the outdoors, some of the very best in the world. And you pay for it. The clothes are so expensive they are referred to as "PataGucci." The story the company tells is as appealing as the high quality of its products: *If you wear this jacket you'll be in the company of great mountaineers.* You aren't just paying top dollar for a fleece pull-over; you're paying to belong to an exclusive club. This is both modern and timeless.

My friends know how much I enjoy my bullwhip by how frequently I break it out. I love the crack of the whip. It's dangerous, provocative. A nice crack grabs attention. I especially appreciate that a good crack is not leather popping on itself. It's a small sonic boom right in the palm of the hand.

There is a certain power in knowing that with just the right flick of the forearm, the tip of the whip will travel faster than the speed of sound. The bullwhip is a fun little prop I like to use for both target practice and cocktail party entertainment. The consequences of failure are painful. While there are a few precautions, the skill in handling the bullwhip comes with knowing it's full danger, if you aren't careful. However, when one understands the flow and flick, the wisp and whim of the whip, it can become an artful dance of sonic booms, a snake of leather swirling through the air as though it were wind. *Crack.*

This is what good messaging and advertising should feel like. A good sentence or a simple tag line should crackle with the intensity of a bullwhip and create its own sonic boom.

Disneyland boldly claims to be "The Happiest Place on Earth."

Nike says *Just Do It!*

The glory of five interlocked rings of the Olympic brand.

The simple word, *Hope*, to capture the essence of a presidential campaign.

A fluttering butterfly.

16

Good as Gold

Omar Jabara and I first met in our early days on a U.S. Senate campaign in the late '90s and have been good friends ever since. For nearly 15 years he served as the head of corporate communications for the largest producer of gold in the world. Every time I talk to Omar, I'm reminded that there is probably no better example of how the art of *story* creates value — real monetary value — than gold.

We were sitting by the Eagle River on a summer afternoon when I asked Omar, *What is so fascinating about gold? How has it held value for so many cultures for so long?*

He explained that gold represents archetypal values going back to the Sumerians and no doubt beyond. "It's shiny and lustrous. Gold is somewhat rare. It's analogous to the golden sun. Gold, with its history of being valued across civilizations, connects us to the past."

Omar said the question as to how gold has retained its value for time immemorial is a complex one. Humans adapted differently than did chimpanzees. Somewhere along the way, we managed to derive general principles from specific scenarios. This ability to abstract is what distinguishes us from primates and other animals. In reality, gold offers little of intrinsic value. It's not like wheat or oil, which offers sustenance and fuel, respectively. Gold's value is tied up in the meaning it has for us in stories that span history — buried with King Tut, worn by emperors and queens, shipped in bullion across oceans in galleons of conquistadors. For us, gold signifies legend and conquest. For chimps, it means nothing.

Omar mentioned that before we developed our current understanding of the way the world functions, there was alchemy, a medieval precursor to chemistry, physics, and geology. Whereas science is rooted in facts and predictable outcomes, alchemy was essentially about creating *belief through story.* Before there was astronomy, there was astrology. Before modern science, there were ancient origin myths, etiological tales, and superstitions that tried to make sense of this befuddling world. Throughout the eras, one thing has stayed the same, however, and that's gold. It still holds our fascination. As Omar observed,

> Throughout history, civilizations needed to assign value to things, and across time and cultures, gold kept appearing as the ultimate store of value. It doesn't lose its value precisely because people believe in it. The more it appeared, the more accepted and valuable gold became. These older things have archetypal properties that speak to us on a deeper level.

Omar referenced Bitcoin sending panic throughout the gold and silver industry in 2015, when some financial analysts predicted that the internet-based currency would undercut precious metals. They believed that Bitcoin would retain value and, thus, render gold obsolete. So far, just the opposite has turned out to be the case. Omar theorized that the advent of Bitcoin actually increased the value of gold because it made people realize they still needed something tangible, something physical to hold onto. Bitcoin exists in the ether. It isn't tethered to anything real — it doesn't have any actual substance, it doesn't occupy any space. Bitcoin, like gold, is only as valuable as people believe it to be. Except we can touch and feel gold between our fingers. But, as any economist would tell you, such valuation is open to a lot of speculation.

This is where the World Gold Council comes in. As the marketing arm for the gold industry, it keeps the narrative of gold alive and well. Gold companies don't talk too much about the production of gold, the messiness of mining. They mostly communicate in the parlance of investor relations. The World Gold Council tells the story of gold; the companies' job is simply to produce it.

POWER OF METAPHOR

Metaphors are ways of making people care about something by relating it to something else, something they can easily understand. Omar told the story of how gold has as much symbolic value — or metaphorical value — as actual value. One could say there's a fair share of subjective reality packed into the precious metal. Omar illustrated the importance of the tangible. And isn't this what metaphor does? It puts an abstract concept in concrete terms or finds an analogy we can all visualize or relate to in some way. As you may recall, I had built my whole speech and debate career on the back of Sisyphus pushing his rock. It was a way for me to simplify whatever policy issue was at stake. In Part 1, I described ways in which we understand and make sense of the world. To understand concepts and meaning, we rely on story and metaphor.

In 1980, George Lakoff and Mark Johnson published a groundbreaking book called *Metaphors We Live By*. In it, they argue that human thought processes are largely metaphorical. They use a simple example — the phrase, TIME IS MONEY — to describe how we think about work in modern Western culture. This metaphor is relatively new. Historically, this was not how people thought about time. They thought about moons, seasons, and harvests. Today, while time is certainly not money in the literal sense, we can trade and quantify time as if it were currency. We measure time in numerous ways — hourly wages, budgets, interest on loans, and billable hours. If you've been convicted of a crime, you *repay your debt to society* by "serving time." Yes, time is a commodity, a limited resource, valuable to everyone, but to some more than to others. There are only so many hours in the day and only so many days in the average human lifespan. Today, as Lakoff and Johnson explain, "We *conceive* of time that way." We now think of time in terms of maximum value and return on investment. Otherwise, it is squandered, wasted on something that did not bring value.

As Dr. Cuddy demonstrates in her popular Ted Talk, metaphor accounts for some of our basic physical traits as well. For instance, a slouching posture and hunched shoulders (like someone using a cell phone) generally indicate sadness or depression, a focus inward. Sitting up straight with shoulders square generally suggests a more engaged, attentive, and positive state of mind. Simple orientation relies on metaphors so ingrained they have become almost second nature. Think about how important the metaphors, UP and DOWN, are. Look *up* to the heavens or *down* in the dirt. He *rose* from the ashes. That person is such a *downer*. The *height* of physical prowess. An athlete's momentum is *up* or *down*. He will *rise* to the *top* or *fall* to last place. Another common dichotomy for metaphors is WIDE and NARROW. Lakoff writes, "Happy is wide; Sad is narrow." These metaphorical constructions help convey the abstract — the emotional, social, psychological — by communicating them in terms of the physical.

Science, philosophy, and common sense are of one accord: we can only come to understand something new by conceiving of it in terms and principles we already know. Lakoff writes, "We can only understand what the neural pathways and networks will allow us to understand." A supremely important fact in formulating any advertisement or messaging. As such, metaphor may be as effective as storytelling in bringing clarity and comprehension to a situation and, ultimately, in controlling our environment.

When you're promoting anything — whether it's selling toothpaste or sparking public interest in free universal child care — it's crucial to educate your audience about what it means to them, and that entails ascertaining what it *has meant* to them in the past (regardless of them having realized it or not) and what it *will mean* to them in the future. How do they fit into the Hero's Cycle? People can't comprehend, and thus reject, communications they do not have the capacity to understand. An analogy is only as strong as the meaning it conveys to an audience. A joke about Newtonian physics usually falls flat for someone who's just trying to find their next meal.

Gold is only as good as our belief in it.

17

Data Is the New Bacon

I was enjoying time with my family on the beach in Sayulita, Mexico, when I ran into Eric Roza, founder of Datalogix. Our families were on vacation separately and just happened to see each other in front of Don Pedro's. We had played music together before and our daughters went to school together and were playmates, so it was a pleasant surprise. I asked Eric if he was interested in supporting me on a swim the next day to San Pancho, but he said he had to take an important call and would be working most of the day. I made a little joke about him working on vacation, and he laughed it off.

The call was with Larry Ellison, the founder of Oracle, who is worth about $64 billion and was listed at the time by *Forbes* as the fifth wealthiest person in the United States. Ellison had built an empire on his understanding of databases and the ever-growing need to manage them. He was on the phone with Eric because Datalogix understood Big Data.

Datalogix was unique in that it married retail sales from catalogs and retail stores with aggregated data from online behavior. Eric and his team found that companies were spending tens of millions of dollars on digital advertising with scant confirmation that it was actually working. Datalogix would match Proctor and Gamble, Home Depot, and other large retailers with digital media and social media to provide such critical information. By coordinating retail consumer data with online profiling, these companies spend less on advertising that is far more effective.

It works like this: an in-store retail purchase drives an ad that pops up on the consumer's phone. Say you bought early pregnancy clothes from Target. They know you are probably going to have a baby in five to seven months, so they can start sending targeted ads to your Instagram account at specific intervals throughout the pregnancy and after. They know you'll need diapers, newborn clothes, that fancy stroller, and all the other must-have baby items. They'll know when a third birthday rolls around, and they'll know from previous purchases if it's a boy or girl for more targeted ads.

On the beach that day in Sayulita, Eric was wearing a shirt that read, "Data is the New Bacon." It meant that decisions should be informed by data. Like bacon, data makes everything taste better.

What Datalogix enables companies to do well is *targeting*. Targeting defines not only the audience, but also the message. Good targeting yields valuable feedback about what your message actually should be, based on the specific demographics you want to reach and what kind of language will appeal most to that target audience.

This is why Larry Ellison wanted to talk to Eric that day about Oracle buying Datalogix. Not long after we returned from holiday in Mexico, Eric was profiled on the cover of the *Denver Business Journal*. The headline read: "This Man Built a Unicorn" (a start-up company that is valued at over a billion dollars). Turns out, it was worth taking that call in Sayulita. Eric sold Datalogix to Oracle for $1.2 billion. The night of the call we celebrated by playing guitar and drinking tequila in Sayulita's town square. Eric confided, "It was the call I've been waiting for my whole life."

With transactions such as these, we now operate in the era of Big Targeting. Driven by Big Data and Bigger Money.

Even bigger than *Bacon*.

Incidentally, Eric is also an avid CrossFit athlete and the owner of the CrossFit Sanitas franchise. In Part 1, we showed several examples of what not to do in crisis incidents, especially those involving race. During the COVID-19 pandemic and racial justice movement spurred by the murder of George Floyd, Greg Glassman, the head of CrossFit, told staff on a Zoom call, "We're not mourning for George Floyd — I don't think any of us are." He spouted wild conspiracy theories about Floyd's death and the coronavirus. He made his insensitive remarks as protests spread in nearly every major American city. After the call, staff and franchise owners were outraged.

A copy of Glassman's rant was leaked to the media. Hours later, CrossFit released the news of Glassman's "retirement." The statement called his comments "incredibly insensitive and hurtful." In an odd twist, CrossFit added that Glassman shouldn't be judged solely by his comments. "Greg believes in equality," the statement read. "He does make mistakes, but he has done more than anyone for this community and created unimagined opportunities for others."

A big misstep. In an apology, never let a defense of your moral character overshadow your contrition.

Either he had media people around him giving him bad advice or he didn't listen to the good advice. His apology was only a half-apology. *Yes, he's sorry, but he's still done more good than ANYONE!* This is why Eric Roza was able to swoop in and buy the entire franchise — previously valued at $4 billion with 13,000 gyms in 158 countries — for "undisclosed terms." Code for pennies on the dollar.

Like so many of the incidents previously mentioned, this one exploded because someone communicated in a way that offended and hurt people. These weren't just misplaced words, they were self-destructive because they exposed an entire belief system of *uncaring*. After Eric completed the purchase, his subsequent letter to the CrossFit community hit the mark. "My view is simple: Racism and sexism are abhorrent and will not be tolerated in CrossFit. We open our arms to everyone, and I will be working hard to rebuild bridges and those whose trust we have lost." He continued,

> I come to you with deep humility and the realization that we have hard work to do. I am committed to listening, I am committed to learning, and I am committed to leading positive change. Most of all, I am committed to CrossFit and to you, as a member of our community.

He concluded, "We are going to crush it, together."

DRIVING PUBLIC OPINION

This subject of divining public opinion — by polling, focus groups, surveys, and consumer research firms — has garnered an immense amount of attention. And money. Statistics is a field of mathematical analysis that quantifies public opinion into data. Data collection has become an enormous industry. Statistics and data collection, which constitute the basis for the whole profession of public opinion research, provide intelligence used to craft advertising and marketing campaigns.

For instance, an old friend of mine, Kingsley Stoken, was an early investor in Ticketfly. The beauty (and perhaps the horror) of Ticketfly is that users created an individual profile in order to buy tickets to events. When Ticketmaster purchased Ticketfly for nearly half a billion dollars, it wasn't the ticketing they were after, it was the data collection and vast troves of consumer preferences that the company had amassed. Ticketfly was a data goldmine.

Nowadays if those who communicate a message use the power of consumer behavior tracking, they have much more certainty that recipients share certain values. If someone buys a tent at the large outdoor retailer, REI, then you could infer that that person might also need a camp stove and a tent. The National Park Service might use that data to entice the purchaser of the tent to come visit a park. Or the city of Moab might target the person with recreational opportunities like boating, mountain biking, or hiking beautiful arches.

There are countless creative ways to target people, but for a moment let's focus on one. Consider fashion as a political targeting tool. An article by Vanessa Friedman and Jonah Engel Bromwich in the Style section of the Sunday edition of *The New York Times* reported on how Steve Bannon hired the British political consulting firm Cambridge Analytica to use fashion tastes and clothing preferences to help build the decisive voting bloc in the 2016 presidential election.

In 2018, Christopher Wylie, the data consultant and whistle-blower from Cambridge Analytica, spoke at a conference organized by the fashion industry. Wylie described how the firm compiled voter profiles based on the clothing choices drawn from data culled from Facebook. Looking ever the fashionista himself, in dyed green hair, a modern fitted gray suit, and a colorful t-shirt, he pulled no punches in his presentation, "Fashion data was used to build AI models to help Steve Bannon build his insurgency and build the alt-right."

Wylie described how many of the great American brands are founded on the myths of the West — frontier narratives featuring a strong, independent male figure. It makes sense that those who buy Wrangler jeans, with cowboy rope font on its logo, would tend to lean Republican. Other brands have other leanings. North Face, for instance, tends toward a more environmental, liberal ethos. "Fashion brands are really useful in producing algorithms to find out how people think and how they feel," explained Wylie at the Business of Fashion conference.

It's not just fashion, of course. It's all aspects of our lives. Who would be surprised to hear that a regular Walmart shopper would be more likely to vote Republican than would a regular Whole Foods shopper? Cambridge Analytica, in its research for the 2016 Trump campaign, determined that a preference for U.S.-made cars was a strong indicator of a Trump voter.

But there is an ethical problem with how Cambridge Analytica came up with this data. *The New York Times* determined that the firm "preyed" on people with its algorithm, which analyzed the Facebook profiles of more than 50 million users without their permission. *The Times* authors wrote, "The event was just another example of how personal data, given incrementally to products and platforms over years, can be used to manipulate individuals in unanticipated and potentially damaging ways." Wylie's talk was ultimately a condemnation of Facebook's enormous power. He described the company as a scourge on society, one that deliberately sows dissension among users based on cultural and political preferences.

These days data collection — microtargeting, profiling, universe gathering — is a double-edged sword, and which edge is sharper depends on where you stand. Communicators have unprecedented access to their audience, but the process of reaching them right where they live raises objections even among the data gatherers themselves and may violate the rights to privacy that Americans hold most dear.

As Dr. Andi O'Conor says, the reason to communicate is to create relationships. These relationships between people who communicate are fundamentally *spatial* relationships. Traditionally, people only talked to others one social level away. At the beginning of human communication, it was critical to be in direct contact with the people you were trying to influence. The king had to directly address his subjects. Today, it's still traditional for the Pope to address the faithful in St. Peter's Square in the Vatican.

Spatial relationships include age, gender, race, values, and other elements to differentiate people, cultures, and organizations. This forms our social structure. It is this structure that defines communication channels. Senior citizens rarely talk to high schoolers outside of a familial or teacher-student relationship. Rich people rarely talk to inner-city kids. Generally, people tend to talk to those of the same station, those who share a certain understanding and set of basic concerns.

Even in the world of modern communications where we have become somewhat removed from the direct address, nothing can move people like a direct personal connection. Seeing something on TV can raise suspicions of fakery and acting. Put someone in direct contact with the experience and it becomes a greater reality. If you have your doubts, go see a live music show. This is why political rallies and public rituals are so important. The original targeting. Basic analog connection.

Sometimes in creating audiences it can be helpful to identify avatars. What does the ideal member of our target audience look like? Present this hypothetical individual with your ideas and carefully envision the reaction.

Just remember, while we may select words or notes based upon the predicted response of an intended receiver, we also have liberty, like a great jazz musician, to play what we want. In the marketplace of ideas and emotions, we are free to throw whatever we want out there and people are free to react in whichever way they choose. We also have the freedom to keep the music entirely to ourselves.

INDIVIDUAL, INTERPERSONAL, GROUP

For thousands of years, nomads roamed the earth and created paths and trails. With excess agricultural goods to trade, prosperous communities linked with one another and routes were established. Holy sites attracted the faithful. People made years-long pilgrimages to the Vatican, Mecca, Kyoto, and Santiago de Compostela in Spain, to name just a few. They traveled long distances on foot and then, after reaching their destination, performing their rituals, doing their penance, and recovering, they would have to walk back.

As the world modernized and transportation became easier, from steamboats traveling upriver to railroads cutting across continents, so too did communication. The space that people traveled and were able to communicate within kept expanding. Today we can reach out to damn near anybody. As communications scholar, Karl Rosengren, states, transportation and communication have always been "mutually interdependent and complementary." Wherever there are people, there is a need for both.

Let's distinguish among the different levels of interaction. There is the *Individual* (the constant conversation you have with yourself), the *Interpersonal* (the conversation that happens between individuals), and *Group Communication* (the conversations that happen among groups, organizations, and institutions). The interesting dynamic is that the latter two inform the first. The web of connections between these levels is a testament to the complexity of our human consciousness. That which distinguishes us from other living organisms on the planet.

All organizations die at some point or another. Rome fell. It was once said that the sun never set on Spain, and then it was England whose empire spanned the world. Those "empires" now belong to the past. Banks and corporations that were "too-big-to-fail" have failed. But organizations, groups, communities, tribes, unions, and families — whatever you want to call a collection of people — are still the very root of how we interact. Rosengren writes, "Communication may thus be regarded as a defining factor of human groups and organizations of all degrees of size and complexity." Yuvall Harari maintains that this ability to think outside ourselves was the key to *Homo Sapiens* dominance even over similar species that were physically stronger.

Polling and focus groups can be another way to narrow your target. Around election season, many tend to think of polls as being useful mainly for prediction. But elections have shown us that we can't predict everything, so what's the use of polling? What polls provide is a snapshot of people's moods. In some sense, polls constitute a record of how people *felt* at a certain time. Perhaps after a major event or about a divisive person or issue. Recall how Tommy G. wanted to tap into the current *zeitgeist* with his butterfly — polls can be a means to do just that. If we can predict a trend, we can ride it or reject it as we see fit.

TRIBALISM

When I lived in Telluride, Colorado, in the early nineties, I was fortunate to work for the futurist John Naisbitt, author of the bestselling book, *Megatrends*. He hired me as a research assistant on his book, *Global Paradox: The Bigger the World Economy, the More Powerful its Smallest Players.*

He devoted an entire chapter to the rise of *Tribalism*. He explained that we have a natural affinity for organizing in all kinds of groups: clans, clubs, teams, etc. Hunters usually socialize with hunters. Sea-faring folks with fellow shipmates. Masons with other masons. Coders with other coders. Fraternity boys tend to date sorority girls. It's natural. The literature argues that self-segregation usually starts in high school, when students begin to develop a sense of identity that incorporates the history of their own particular demographic makeup. There is nothing inherently wrong with this natural grouping.

John Naisbitt wasn't the first to explain the concept of tribalism as people organizing along lines of a common identity or set of values. This is not necessarily a good or bad thing, *per se*. But if the tribe's attachment to its shared identity creates tension or violence, an inability to find common ground, or an "us" vs. "them" mentality, it can be divisive and dangerous. Naisbitt predicted the dark side of Global Tribalism: instead of putting the entire society first, people value the primacy of their tribe over all others. In fact, the Great American Experiment is about the collaboration between tribes. It might be an uphill battle. Many scholars believe evolution happened in small groups and not in mass societies. But if people constantly put themselves and their tribe above the whole, then anomie sets in and the whole fragments.

The intention of serving only the tribe over the good of the whole is ultimately unsustainable. This kind of behavior increases polarization among groups and exacerbates their inability to relate to or understand each other.

Communication cancer.

18

Our Digital World

At this stage in our collective tech dependence, it's hard to believe that when I was growing up my only connection to the world at large consisted of a corded telephone on the wall, a transistor radio, and a television with five channels (three broadcast networks, a public broadcasting station, and a Christian channel). I didn't use a computer until after college. When I did research for speech and debate competitions, I had to search for printed publications on microfiche at the library. When I worked at Commander's Palace, I took all the reservations over the phone and jotted them down in a big book with a ballpoint pen. We didn't have cell phones. When the internet emerged, it seemed like a novelty.

But there were those who foresaw the dawning of a new age of telecommunications. In a world of VHS rentals and clunky answering machines, these visionaries were thinking post-analog. All the content you could ever want right at your fingertips. I now wonder if these farsighted ones who anticipated the launch into the digital age knew that it would be equal parts exhilarating and exasperating.

While there is much truth to the notion that our world is more connected and convenient than ever, it is also true that not all the effects are positive. Much of Main Street is crumbling in the face of competition from the online behemoths who deliver products to your front door lickety-split. Automation is taking jobs from workers unprepared for the technological leaps and bounds. Millennials surf the net on laptops at coffee shops and hostels looking for the next fix in the gig economy. We enjoy instant information about most everything, but, as Jennifer from Excel Energy reminded us, it can be argued that cat videos have left a whole swath of the population listless, complacent, and distracted. Then there are security breaches, personal identity theft, hacking, cyberterrorism, and artificial intelligence. As Andrew Wylie from Cambridge Analytica warned us, the 494-page report Facebook delivered to Congress reveals just how much the social media titan knows about us.

My generation, with still-vivid memories of phone booths, busy signals, and dial-up internet, understands just how paradigmatic this societal shift has been. My kids, on the other hand, will never know what it is like *not* to have smartphones and uninterrupted hyperconnectivity.

Social media offers us a tremendous opportunity to connect in ways previously unimaginable, but it also comes with a unique set of challenges. The great paradox about our digital age is: *As we become more connected, we also become more isolated.* It often seems like we're breaking down as many social connections as we're building up, substituting online communication for the real thing.

Aaron Sorkin, who wrote *A Few Good Men, The West Wing,* and *Moneyball* in addition to many other highly acclaimed films and TV shows, told *The New York Times Magazine,*

> The problem I had when I wrote *The Social Network* was that this thing that's supposed to bring us closer together is pushing us further apart. It gives everyone the impression that everyone else in the world is having a better time, and that if you are not cataloging your life, then you're not really living it. People are going to show you only pictures of themselves having a great time at the best party with the coolest people.

He continued,

> When we're a little kid on a playground and say something mean to another little kid, we see in their face what we did, and we feel bad because of it. On social media, it's more like yelling at another driver from your car.

A deeper understanding of social media in today's environment demands an acknowledgment of the generational divide that it so often evokes. It is indisputable that modern hyperconnectivity is having a profound impact on how today's youth and millennials relate to the world. Thanks to a wide variety of social media and video chat apps, phones are now intimate parts of all relationships, but especially those of young people. Whereas in my day, teenage couples met at high school football games, today they meet through avatars in virtual reality games. Of course, it's easy to cast a critical eye on the new and unfamiliar; for time immemorial every generation has decried the one that succeeds it.

To find out more about technology's impact on how teens communicate and to distinguish legitimate fears from hidebound technophobia, I turned to author, educator, social media expert, and good friend Rosalind Wiseman.

Even if you don't know Rosalind Wiseman, you are probably aware of her influence. In 2002, Tina Fey read the *Sunday New York Times Magazine* cover story about Wiseman's book on teenage girls' behavior, *Queen Bees and Wannabes*. The book inspired Fey to write (and co-star in) *Mean Girls*, a film that would go on to achieve cult status and define the high school experience of the Generation Y set in the early 2000s.

After Wiseman published her groundbreaking book, she wanted to better understand how we, and especially young people, interact in the world of social media. Why, with so many of them participating in the massive social networks, do so many young people feel so disconnected? A 2019 YouGov poll found that 22% of millennials report having no friends. More than a quarter of all teens are affected by some form of depression. This generation is also online more than any other. In 2015, a Pew survey discovered that 73% of teens had a smartphone or, at least, access to one. By 2018, that number had shot up to 95%. In 2015, 23% of teens said they were "almost constantly online;" in 2018, it was 45% with YouTube, Instagram, Snapchat, and Facebook ranking as the most popular social media platforms.

With these platforms and devices, there has been a dramatic rise in cutting, eating disorders, opioid addiction, and general mental health disorders in the U.S., particularly among teens. There are more prescriptions for Ritalin and Adderall to address attention deficit disorders than ever before. School shootings are now commonplace. There is a much greater emphasis on testing in schools, which increases stress for students and teachers, who are both judged on test performance. These are not mere coincidences. They are symptoms of the intense environment in which our children are growing up. I helped AT&T launch their Believe effort at addressing youth mental wellness. My work with education associations and in juvenile justice reform has shown just how tough it is for kids today. From my work with educators and teachers, I gather that mental health issues pose some of the biggest challenges in public education — and are some of the most underfunded.

I first started a conversation with Rosalind over drinks in Austin at South by Southwest after she spoke about online bullying in one of the festival's most well-attended keynote talks. A few months later we resumed the conversation in Boulder, CO, where we both live.

Rosalind reminded me of Marshall Ganz when she identified the two stories we tell: the one we tell ourselves — our personal narrative — and the one we tell others, which she describes as *The story we want others to believe about us.* This is especially common among kids, who may present their public face, or "brand," on Instagram and their private self on Snapchat. These two accounts may constitute completely different personas. Rosalind elaborated, "So there is always this tension between social status and real story, between the public brand and the real self. It is always grinding and it's increasing their anxiety." This tension between competing identities precedes the internet — just look at politics — but she contends that social media has exacerbated it.

Rosalind pointed out that most adults are hypocritical when it comes to screen management. Many parents send mixed signals: they expect the kids to limit their cell phone usage while they themselves use their phones at the dinner table. Technology addiction, which affects all generations, not just teens, is a serious concern. She believes that sustained device use causes a physical change that reflects one's internal state. Prolonged screen time affects posture, causing the shoulders to slump, body language that reads as passive and disaffected. As Dr. Amy Cuddy explained earlier, such posture suggests disengagement with one's surrounding environment. One's device gets all the attention.

Rosalind and I both lamented what she calls "the tyranny of unlimited options." There are too many choices. With so many options, it seems more difficult than ever to make plans and stick with them. These days the satisfaction of making a decision pales in comparison to the dissatisfaction of *not* making alternative choices or, rather, not being able to make *all* choices.

For much of this we can only blame ourselves. We eagerly participate in an online culture that has produced influencer marketing, the attention economy, and associated terms, such as *thirst trap, humblebrag,* and *hate-like.*

Data might be the new bacon, but think of all the fat rendered from a plate of crisp rashers — it can be too much of a good thing. We can attribute responsibility for these all-consuming online worlds we now inhabit to those who created them. It seems they set the parameters in such a way as to maximize profits over privacy and social well-being. We are not just the consumers or users, we are now *products.* Our personalities, social lives, and aspirations have become algorithms and data sets for profit.

Is it any wonder that people — and particularly the young adults that Rosalind Wiseman works with — feel so cynical, as though they're always being used for someone else's selfish gain? Perhaps they feel like they're trapped in an endless labyrinth or on a hamster wheel desperately searching for meaning. Ours is a data-driven world. Unless you choose to live off the grid like a hermit — which I completely understand — there is a need to live with today's technology. The unfortunate reality is that so many of us feel we must accommodate our lives to technology when it should be the opposite.

Let's be practical, the reality is technology works in today's working environment. Putting aside all these issues: if you want to reach an audience, any effort at communication should factor in a digital component. This allows for better metrics and better data at a fraction of the cost of most other advertising with more precise targeting. In terms of efficiency, when producing a TV ad, hanging a banner, or standing on a street corner, it's difficult to know just how many people are paying attention. When it comes to direct mail, for instance, how many people toss out campaign mailers without giving it a second glance?

I would be remiss if I didn't mention that Rosalind, who has charted the darkest trenches of social media, still has hope. She is skeptical about an across-the-board "communication breakdown" in kids and society in general.

> Most kids want to address the problems they see in the world. They want to have a sense of purpose in their lives and a way to contribute. The trick is making the information we share with them relevant and respectful of the fact that it's really different to grow up today.

She believes that as long as you show them a path and explain *why* something will be better for their friends and community, then young people will be motivated. She has witnessed firsthand how, not just kids, but everyone is hungry for meaning in their lives.

In a phrase that pertains to so much in this book — from silkworms to butterflies to the World Gold Council — Rosalind concluded, "If it gives them meaning, they will do it."

19

The Holy Trinity

What I love most is *getting it done*. I'm talking about the tactics and the disciplined act of bringing all the elements together to make a plan actually happen. It's where the rubber meets the road. Completing the planning process and reaching the point when it is time to get out there and fly. The goals have been set, the approach defined, the audience determined, and the messaging baked to a fine soufflé.

The word, *campaign,* is most often associated with political endeavors to win elected office, but contemplate for a moment the innumerable varieties of campaigning. Any time one sets a goal or strives to realize an idea — from the most elemental to the most sophisticated — there should be a campaign around it. There are marketing campaigns for products, such as cosmetics, ice cream, and cars. There are campaigns to win others' hearts, to win a Grammy, to get people to use more renewable energy, to recycle more, to buy American products. In the original sense, there are military campaigns. Campaigns are wonderful laboratories for studying human nature and practicing situational awareness.

Whatever the campaign, the same dynamics and processes are involved. These include storytelling, messaging, targeting, executing tactics, and managing the interplay of time, money, and people. Decision makers of any campaign that involves advertising and public relations must determine how to best utilize these limited resources.

Nowhere is the desire to control public opinion more on display than in campaigns for public office. There's nothing like the thrill of election night when the vortex of an entire political campaign funnels down to a single point: the last bit of breathless anticipation before the final result is announced. Campaign headquarters are a cauldron brimming with intensity. At least, they should be. I'm talking frantic urgency. If you walk into a campaign HQ and the phones aren't ringing off the hook and people aren't running around as if their hair's on fire, you know they're in trouble. A silent campaign headquarters is the sound of defeat.

Because of the intensity, the cauldron stirs its own craziness. A friend once imagined that there were more sex, drugs, and rock and roll on a U.S. presidential campaign than on a Rolling Stones tour. The players just happen to be wearing blue blazers. Campaign endeavors bring out the best in people — and the worst. They are supreme tests of a person and what one holds most dear.

That's why I love them.

Campaigns mean something. I may be ostracized for saying this, but it doesn't really matter which football team wins on Sunday. I love the New Orleans Saints, but ultimately their victory or loss doesn't mean much more than the civic pride it brings to the fans and perhaps the economic engine it provides in those fleeting moments of glory.

Political campaigns, battles of values and beliefs embodied in one person, are the most intense, conflict-ridden enterprise anyone can engage in — besides actual war (which, incidentally, often results from contested political campaigns). That's probably why Professor Marshall Ganz, who worked on a number of campaigns himself, uses the classic dramatic structure to describe one. After getting involved with the Student Nonviolent Coordinating Committee and participating in 1964s Freedom Summer, Ganz signed up with the legendary Cesar Chavez to organize California agricultural workers. He spent 16 years with the United Farm Workers and became its director of organizing. I appreciate Ganz's reflections on the energy campaign workers and volunteers draw from collaborating with one another. I especially appreciate his articulation of a sense I've often had when working on a campaign, that I'm involved in a grand narrative with a distinct arc.

The course of all campaigns — military, advertising, and issue-based — can be likened to the unfolding of a story. At first, there's a period of education and introduction (prologue), then it kicks off with a bang (curtain goes up), builds momentum toward successive summits and valleys (Acts One and Two). The effort culminates in a final peak of critical decision making (climax). As the results are revealed, all is reconciled and resolved as we celebrate the outcome (denouement). Our efforts generate momentum, not in a vacuum, but as a snowball.

As we've previously discussed, what people care about in all their interpersonal experiences — be it a romantic relationship, friendship,

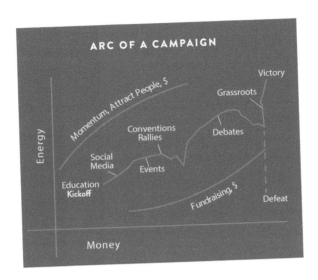

FIGURE 19.1
Arc of a campaign.

acquaintanceship, or random run-in — is how they *feel*. And it's the same for billionaires, baristas, and beggars. The brain responds to feelings, not facts. Taking this to its logical conclusion, one might even be tempted to say that facts are only as important as their ability to elicit emotions.

One could say the same about campaigns, those of any stripe. Policy might be the flesh and bones of the body politic, but emotion is the heart and soul. Every election season, the drama of humanity is staged yet again. Part psychology, part organizational management, campaigns are vessels for our hopes, dreams, beliefs, and values.

TIME, MONEY, AND PEOPLE

At Commander's Palace, I learned why onions, peppers, and celery are regarded as "the Holy Trinity" in French and Creole cooking. These three simple elements, when chopped and sauteed together, become greater than they would alone suggest. When cooked down to a soft reduction, they form the basis of all Creole cooking. Similarly, Time, Money, and People form the foundation of campaigns — and any endeavor for that matter.

Tactical elements don't appear by themselves; one must *produce* them.

Let's recall the First Law of Thermodynamics from our coffee talk with astrophysicist Dr. Jeff Bennet in Chapter 1. Bringing order out of chaos requires energy. Fledgling ideas that break forth from the wall of information noise require attention. Campaigns require resources.

And there are three, and only three, golden founts from which to fill the cup. *Time, Money, and People.*

My good friend, Mike Antonucci (Nooch), is a producer at Disney Corporation in Los Angeles. It's an intense job, for sure, one that comes with a lot of responsibility and pressure. We have gone on many adventures together and a number of trips to New Orleans for the Jazz and Heritage Festival. He has produced numerous world-class commercials and campaigns for Disneyland, Kellogg's campaign for the Sochi Olympics, Chevy Global, Budweiser, and Busch, just to name a few. He was on the team that put the cheese before the macaroni with Kraft Cheese and Shells. (A subtle change of *index* as the emphasis is now on the cheese, not the noodle.) Nooch can create anything — with the right budget. Producers literally create an environment from scratch. Sometimes using parts of the natural environment, but frequently creating realities out of whole cloth on soundstages. They are masters of capturing a perception of reality.

The process of making a spot is such that an art director and writer come up with the idea and concept and bring it to the client for approval. A producer then takes the script from the art director to create the spot. There is a script, a deadline, and a budget. But so far everything is only on paper. Producing, in Nooch's words, is "striking a delicate balance and always figuring out a way to get it done on time and on budget. Pulling ideas out from the ether and giving them shape in the real world."

Nooch explained his general approach to me. When he produces a spot, his first step is to come up with an idea that will lead to the purchase of the product. His second step is to communicate that idea. Neither step is the most difficult part of his job. For Nooch, the art isn't creating the spot, it's managing all the different stakeholders. Yes, he has to produce it on a budget within a constantly shrinking timeframe — and that's tough enough — but he's also expected to manage the talent, the client, the lighting guys, the physical location, the works.

The problem is that the story has a different personal meaning for everyone involved in the project. The key to realizing a creative vision is to make sure everyone understands what they are creating, even if every time someone touches the product — from writers, clients, music composers, artists, and actors — they leave their own fingerprint on it. "Why and how we create the environment of the commercial is extremely important." He is constantly asking questions such as, *Why is something in the shot? Why is something not in the shot? Why use this person rather than that one? What is the setting in time and place? What are they wearing and why?*

The way Nooch describes it, "It's the producer's job to make sure everybody understands the same vision. Or, if they have different points of view, to still keep the integrity of the piece." The most fundamental part of his job is the communications involved in uniting the crew around a single vision. And there is also the financial aspect. There are compelling options for every little decision, but how much will each cost? If the basic idea changes in the middle of the creative process, costs can skyrocket since production companies base their bids on projected shooting schedules. The lawyers and executives think it should be shot one way, but the director has another idea. Nooch is in the middle, drawn and quartered.

If there is a change and someone says, "I think this calls for a helicopter shot." Nooch's job is to say, "No. This doesn't call for a helicopter shot." But, and this is crucial, he may follow up with "*Why* do you think we need a helicopter shot? Is the current establishing shot lacking something? What? What are we solving for? Can we address the issue within budget?" Nooch's sage counsel is "*Focus on the what, instead of the how.*"

Or let's say a critical shot gets postponed due to weather and now an eight-hour day becomes a 14-hour day. Everyone still needs to be paid. So, who eats the extra cost? The ad company, the client, or the production company? Ahh. Nooch must make choices. The dance of production rolls on.

Nooch compares his work to a home remodel. It always takes longer and costs more than you think.

He says the tough part about being a producer, as in much of life is, "The ever-present fight to be creative and also financially responsible with the resources you have. That is the tension in being a producer."

He describes the whole creative collaborative process. "Every time someone contributes, it evolves." The actor, the writer, the camera guy, and most importantly the executives have a lot of opinions. As the producer, you must manage these interests to make the vision a reality. The magic is how a producer assembles a team of various independent artists to execute a grand and complex vision. The most important part of the job is not actually *producing*; it's getting everyone who matters to understand the vision and work toward *realization*.

He believes that more than any other, "Storytelling is the industry of our time." Late one night in the shed in my backyard, he told me, "Back in the Mad Men days it was more conceptual. Very image-oriented. Now it's more retail-oriented. Like *Toyotathon*." He mentioned commercials whose crews weren't quite sure what they were selling. The brand was barely mentioned. "There was no product benefit or reason to care," he exclaimed.

Later at night Nooch ventured into the *Tyranny of Options*. He lamented the inability of people to make solid plans. "Now communication is fragile," he said. "There is more of a sense of always thinking, *What are the cool kids doing?*" Nooch was speaking about his work and the inability of people to commit. As we discussed last chapter, while there is nothing inherently wrong with hyperconnectivity, it is radically changing social landscapes in ways we may not be prepared for.

"An unwillingness to stake out your own ground," Nooch said.

The point being, keep your word and live with integrity. If you say you are going to make an advertisement, or even do something with a friend, then do it. Make plans and do your best to Show Up. On time and ready to go. It's the only way to be an effective producer — not just of advertisements, but of anything.

Nooch was echoing the problem of FOMO. What makes this fear so pernicious is the reality that these days it seems like almost all plans *can* be changed or broken. Nothing is solid. This is a pet peeve of mine. It's also ironic: if you're noncommittal about something so you don't miss out on anything, you end up missing out on just about everything.

The world belongs to the people that produce it. The people who assemble the right pieces and bring to life visions and dreams. From words on the pages of a script to an award-winning commercial or song or play or press conference.

The critical point here is that this interplay is similar across all forms of communication. The dynamic doesn't really differ from the White House to the classroom. People use the same principles across different disciplines. The wonderful part about being a communications professional is that everyone needs help applying these principles.

No part of communications planning is static. As previously mentioned, the OODA Loop is a cycle because it should be updated constantly based on changing conditions or a new understanding of current events. Because of the quantumness of all communication, we can never know for sure how it will turn out. It may turn out exactly as planned, suddenly go viral and catch on, or it just may fall flat for no reason at all. The message, the audience, and the tactics all come together in a cohesive form. They make a gumbo that is your communication strategy.

LAWYERS, GUNS, AND MONEY

Warren Zevon played on the West Steps of the Colorado State Capitol at the rally I organized for the Lisl Auman campaign. He wrote a legendary song called "Lawyers, Guns and Money," which is a clarion expression of the cold realities and limitations of Time, Money, and People. In all campaigns, advertising, political, or military, you come face to face with the all-important matrix of limited *Time* — a deadline, court date, election. This is the gun. While you might be after tangible assets, like a bus or a plane, usually the best resource is cash *Money*. But the most important piece (in the advocacy equation, at least) is *People*. (Lawyers perhaps, but not always.) Nothing happens without people caring. Money is important, but people are usually the most important ingredient in the Holy Trinity.

Time, Money, and People represent the interplay of art and science. Managing these three gets right to the essence of the communication strategy. Deadlines arrive too quickly, the ad-budget runs dry before you know it, and no matter how many people are working, you always need more. Time, money, and people are finite resources for the alchemist and must be used wisely.

This holy trinity also lends parameters and scope to any endeavor. We must ask: *What is the interplay between these three resources? How do we squeeze every last ounce of value out of these magi?* Think of all the available means of communication — television ads, social media, direct mail, phone calls, directly knocking on doors with paid or unpaid canvassers. My counsel has always been to pick a few tactics and do them really well. Even though there is a multitude of ideas out there for engagement and education, *don't try to do everything at once.* People will always come along and ask, *Why don't we do this and that?* Just like someone's always asking Nooch for a helicopter shot.

FIGURE 19.2
Time, money and people.

We have to stay true to the intention and figure out the winning combination. Either we tell people, *That's a good idea! Definitely worth considering...* Or *Get the hell out of here!* We have to be the conductor of an orchestra and bring our three precious resources into symphonic harmony.

Ahh, yes... Let's circle back to strategy. Many of the great scholars (Clauswitz, Liddell-Hart, Machiavelli, Freedman) believe that strategy is ultimately about how to augment power as you strive for an outcome that is greater than the sum of its parts. Ultimately, strategy should frame an identity and purpose, create meaning, and inspire action.

How do you know which is the right tool? In the following example, we will consider where the target audience is currently focusing its attention (spending its time and energy) and identify the appropriate "immediacy" and "intimacy" levels of the communication channels so as to engage them effectively.

PROTECTING WESTERN WATER

Let's use the example of protecting water in the American West to illustrate the interplay of tactics and the decisions that drive them. Our team was called upon to help raise the profile of the Colorado River Water Conservation District, which manages the headwaters of the Colorado River, the most important — and most endangered — river in America.

The river is facing two massive threats: increasing demand and shrinking supply. There is 20% less water in the stream now than just 25 years ago (1995). That's a big problem when 40 million people depend upon it for their livelihoods. And the headwater counties have been identified as the place in the US where temperatures are rising the most — by 4.2°.

As the first person to ever swim the Colorado for 47 miles through Canyonlands, I consider the river close to my heart. I have worked with American Rivers for many years to protect water sources. We advocate for the river system throughout the entire Colorado River basin. So, this particular assignment was very personal for me.

For the entire campaign, roughly half a million dollars was designated for efforts to reach about 300,000 people. Our ballot measure for a small increase polled well at 63% approval for the ballot language in July 2020.

While the entire audience included conservatives, independents, and progressives, one fact unites all the people of the geographic headwaters of the Colorado Rivers: water. This made the messaging fairly simple: Protect West Slope Water. Which also tested at 83% approval when tested.

But as they say in the West, *Whiskey is for drinking. Water is for fighting.*

Our job was to figure out how to tap into the unique *topophilia* of the region, the strong sense of place among people who live in the West. Our target audience possesses a practical, independent spirit. However, these particular voters are very tax-averse. They are a more conservative electorate rooted in the land, agriculture, and recreation. Our unique task here was *Could we make them love their dwindling water more than they hate higher taxes?*

In analyzing our audience, we knew they consisted of older established voters. We drew up an initial universe of "conservatives" and "independents" over 55 who had voted in the last four elections. Out of a possible 300,000 people, this universe totaled 36,000. With our limited resources, we couldn't talk to everyone, so we chose voters who have been "hyper-involved." These are the folks who talk to others at the feed store, at little league games, and over coffee after church. We laid our money down that this unique group of engaged people would be the sprinkler system to irrigate our targeted crop of potential voters.

Our polling indicated that older Republican women were the most valuable (and obtainable) prize. If we could convince them to vote our way, then we might win. By analyzing the cross-tabulations of the poll, we also determined there was big swing vote potential in independent/unaffiliated men ages 35–55. This was an audience that liked to hunt and fish, probably had kids, and was concerned about future generations. We also knew that the real key demographic lay in Mesa County, whose population was twice as large as the next biggest and which constituted the only true media market over a large, sparsely populated geographic region. We had to meet our audience where they were.

These are folks who still listened to the radio, mostly kept up with the local newspaper, and were likely to read a nicely designed piece of mail. While they were not progressive liberal voters, they were users of social media in surprisingly high numbers, likely as a way to keep up with family, especially grandkids.

While we tend to think that our target audience consists mainly of our avatars — the exact profile of the ideal audience member — in actuality, audiences are much more fickle and fleeting. People usually possess a whole assortment of competing beliefs and passions. And most people don't remember more than manageable, bite-sized chunks of information. The reality is that our brains are wired to forget just as they are wired to remember. Perhaps that's why the butterfly worked so well. Anything multisensory, such as moving images, music, touch, and feel, increases retention and understanding. In addition, if there is a call to action, a number, a website, or a drive to mobilize, then people are much more likely to remember the message.

FIGURE 19.3
Protect West Slope Water.

This is how we broke it down:

Direct mail: Five pieces. Target republican women and men over 55 who voted in the four out of four elections and unaffiliated voters under 55 who cared about water and recreation. We figured they would be our most influential group in the top four most populous counties.

Radio: Buy spots in every corner of the district, especially on radio: talk, Christian, and sports. Create three spots. Voiceovers of older male rancher, middle-aged female, and Latino *en Español* for Spanish language radio. Radio is a cornerstone of our advertising strategy because in this region people still drive around a lot and want to stay connected.

Social media: Create a target audience of social media users in the district of 170,000 people matched to the Voter Action Network. Rotate seven vignettes and include a call to action: "Sign up to become a Champion for West Slope Water."

Digital: Include keyword searches on Google and other search engines.

News Media: Consists of three advertisements in every paper in the district, with guest opinion columns and letters to the editor. Meet individually with papers to ask for their editorial support. Establish a regular media relations channel to all the reporters for ongoing announcements of endorsements and other issues as they arise.

The measure won by 72%. A tax increase in one of the most conservative places in America.

I have worked with Matt Rice and American Rivers for many years. They are the most effective advocacy organization for rivers in America. We've been through a number of heated battles together to protect rivers. American Rivers has made many successful films over the years and brought a voice to how we should think about water if we care about making it last.

"The power of our stories is not in us touting our accomplishments," Matt says. "It's in helping allies and partners tell theirs. Those that depend on the rivers tell the stories."

He described how the films and stories have a life way beyond a film. He cited their short film, *Milk and Honey,* in which ecumencial Latino farmworkers share their spiritual connection to rivers. It won numerous awards at film festivals. He also mentioned their movie, *A River's Reckoning*, about a ranching family and its unique approach to water conservation. The film doesn't even mention American Rivers, but it has done more to connect a unique triumvirate of ranching, the river, and people than any press conference ever could.

Matt also described something special about the process,

> Our approach to storytelling is not just about the final product. It's the *making* of these films that leads to deep impactful partnerships. The very act of assembling the story and the people builds relationships and a shared experience. We ride around ranches with farmers and cattlemen. We break bread with Native Americans. We talk late into the night with Latino families about their water. Rivers brings us together.

Matt harkened back to his earlier days in water conservation and activism when it wasn't so. "Back in the day a conservation organization would do a film to showcase their environmental accomplishments," he said. "Then 150 people watch it, maybe it gets in the paper, and it goes largely forgotten."

Thoughtful stories about people and their connections to their rivers have a longer and more impactful life. What Matt Rice and American Rivers have done is to tell stories through videos and images that convey what their work *means*. They produced an excellent video about the Colorado River called *I Am Red* that has received millions of views online. (Keep in mind, a video of an environmental organization's staff standing next to the same river at a press conference would probably get no more than a few hundred hits.) The old way is to focus on policies and accomplishments. Directly addressing the policy objectives (i.e., "We need to save the salmon because they are important."). Many environmental groups and nonprofits still use a guilt-based approach. They make you the villain in the Hero Cycle. With American Rivers, the *River* is the Hero. Through their adroit storytelling, they make you the helper.

There is a bewildering array of many communication channels we can choose from. For this job, we could have chosen billboards, emails and texts. We did television commercials (good, quick, but not targeted). The campaign could have hired people to go door to door (expensive and not practical, given the geography.) We could have chosen the same channels, but adjusted the timing and/or the particular method. But there are finite ways to spend a limited budget.

CHANNELS AND ACTIVITIES

News conferences
Editorial board
meetings at newspapers
Radio talk or call-in shows
A benefit race
Parades
Web links
Conferences
One-on-one meetings
Open houses
Speeches
Hotlines
Listservs
Information Fair
Materials to Support Activities

News releases
Fliers and brochures
Opinion editorials (op-eds)
Letters to the editor
Posters
Public service announcements (PSAs)
Bookmarks
Video presentations
Web pages
A float in a parade
Buttons
Pins
Ribbons
And every other kind of swag.

FIGURE 19.4
Channels and activities.

Consider what kind of product or issue you have, what kind of audience you're targeting, what you want to achieve, and then determine where your audience gets its information. In terms of communication, what are the most direct channels — those invisible wires into a person's heart and mind? Which channels maximize impact? Remember that the method you use says a lot about the message itself.

But these communication channels are rarely free. They almost always require an investment of, at least, one of your precious resources: time, money, and people.

20

Rocking the Vote

Music has long been a powerful tool to influence behavior and give voice to emotions and belief systems. Songs can capture a time and place. A rhythm can capture a romance. Music is a language all of its own. Like the aroma from a kitchen or the pheromones from a silkworm. Think Bob Dylan. Bruce Springsteen. Bob Marley.

Rock the Vote was an organization funded by MTV and the Recording Industry Association of America (RIAA) to get young people involved in the political process. It used musicians and celebrities to encourage young people to vote. Based in Santa Monica, California, it was hip and very successful in mobilizing the youth vote, I served as the national field director of Rock the Vote in the '90s. The job combined two of my favorite things — music and politics — so it should come as no surprise that I loved it. As a team, we organized shows with Neil Young, Bono of U2, and many other of the top artists of the day.

Getting such a gratifying job, however, entailed one of the most bizarre and nerve-wracking job interviews I've ever experienced. My first interview was with Ricki Seidman, Bill Clinton's former Director of Scheduling and Advance, Deputy Communications Director, and Assistant to the President. I flew out to Los Angeles for the interview and arrived at the office on busy Sunset Boulevard to meet her. First, we went next door to Wendy's, where she grabbed an extra-large iced tea. Then she led me to the bus stop on the corner. *Okay... where is this going?* I wondered.

We sat down under the little bus shelter, she sipped her tea, and we made small talk. Then, as if the thought had just sprung to mind, she told me to stand on the bench and pretend I was at Iowa State University at the outset of the presidential primary addressing an audience of young people. She wanted me to give a speech and tell the assemblage of folks waiting for the

#2 bus why they should vote. I got the same panicky feeling I had standing on stage at the national debate championships many years before. But I stood up anyway, cleared my throat, and launched into a plea about how voting is one's civic duty. Sweat was pouring down my face and the traffic on Sunset Boulevard was blaring. Once again, I called on the spirit of Sisyphus to exercise our democratic right and *finally push the rock of the youth vote over the edge...*

Through Rock the Vote, I had *carte blanche* at most of the music venues in Los Angeles, including the House of Blues. For me, that would have been compensation enough. I have always been an aficionado of live music. I grew up across from Hamilton's Club in Lafayette, Louisiana, which *National Geographic* called the birthplace of Zydeco music. Over the years I've had the pleasure of befriending crazy characters like Beat legend and composer, David Amram, bluesman and hoodoo doctor, Coco Robicheaux, and bluesman, Papa Mali. I've gone to ludicrous lengths to see certain concerts. Two plastic bags hold a collection of ticket stubs to every show I've ever attended. Music has been a defining force in my life. An inspiration. A muse. A motivator.

I speak from experience when I say that music is one of the most powerful tactics for moving people. Music is not just a *soundtrack*; it's the *motivation* for specific causes and for broader social movements. Just about every political cause is closely associated with particular music. Think of the traditional gospel song, "We Shall Overcome," during the Civil Rights Movement. Gil Scott-Heron's "The Revolution Will Not be Televised" for the Black Power movement in the '70s. Kendrick Lamar's "Alright" and the Black Lives Matter movement. This is also why brands are so keen on being associated with particular music that touches certain demographics.

When President Bill Clinton would come to Colorado, I would get a call from the White House to do pre-production and advance work for events. I made sure everyone hit their marks, took care of the staff's needs, and courted the press and cajoled them into their allotted space. I was part of the crew that made sure the microphones worked and that a glass of water was placed under the podium (without ice, that is — you don't want POTUS sounding like he's clinking cocktails in the middle of remarks, and you certainly don't want him choking).

At one event for the President, I was asked to advance the press box on the side of the stage. The band Los Lobos opened up for the thousand-plus crowd. They were rocking and had everyone dancing, which is rare for a presidential event. Then they broke out into their cover of "Bertha," one of my favorite Grateful Dead songs. I was standing stage left, just behind the curtain in a dark navy suit with a secret service earpiece and a small microphone wired on my wrist.

As the band built up the tempo, frontman David Hidalgo announced, "Ladies and Gentlemen, the President of the United States!" President Clinton appeared across stage right in black Wayfarer Ray-Bans and wailed on his saxophone. The crowd erupted.

A president on sax playing the Dead. This was my kind of country.

VALIDATORS

Let's imagine you've employed the best tactics and managed to drum up a lot of resources. All that's great, but you still face one of the biggest challenges of communications: planning, deciding, and coordinating *who* will deliver your message. Rock the Vote had tapped into a rich vein of motivating people by using musicians and celebrities as validators to express the importance of voting. The messenger is just as important as the message.

Not all situations call for celebrities, however. Celebrities are also capable of complicating matters. Some situations call for something entirely different. The messenger influences how communication is received. Oftentimes, a spokesperson, the person most closely associated with a crisis, isn't the one most directly affected by it, but the one who speaks about it most effectively, giving the impression that he or she assessed the matter objectively before drawing conclusions. The importance of validators was on full display during the COVID-19 pandemic. Places where scientists and epidemiologists first served as the primary communicators, such as Seattle and San Francisco, were more effective in dealing with the outbreak. In cities where politicians were the spokespeople, where science was secondary to political outcomes, the citizenry didn't trust the information as much and suffered the consequences.

Third-party validators may be academics, doctors, city administrators, scientists — basically any recognized expert on the subject matter in question. When used correctly, validators add potency to a message. The best third-party validator is one who is prepared in advance and committed. The preparation is not to be taken for granted; it can be an involved

◈Ignition

THE IGNITION COMMUNICATIONS TEMPLATE

Analysis & research

Approach

Goals

Objectives

Storytime

Data is the new beacon, targeting

Tactical execution

Strategy

Implementation

Evaluation

FIGURE 20.1

Ignition communications template.

process. Proceed carefully. People do not always say what you want them to say — especially in the heat of the moment with cameras rolling.

If we harken back to the Court of Public Opinion, you'll recall that lawyers are particularly adept at using third-party validators because they often need other people to tell their story, to confirm it. The lesson is that it is much better to have other people talk about you or your cause than you yourself. People who are always speaking about themselves tend to give the impression that they can't find anyone else to do it, and for that reason they're usually doomed.

Validators are the gladiators of communication strategy.

Part 3

Why

It had long since come to my attention that people of accomplishment rarely sat back and let things happen to them. They went out and happened to things.

Leonardo Da Vinci

21

Days of Action

The sun was shining like a beacon over the Rocky Mountains as we gathered at the Colorado State Capitol on a beautiful late April day in 2017. Over 14,000 educators and teachers from all over the state had assembled on the West Steps of the Colorado State Capitol and spilled over into adjacent Civic Center Park. There were chants and creative, funny signs — "Teachers just wanna have FUNDS."

The struggle to strengthen public education and keep students from falling into the abyss of ignorance is never-ending. At the rally, one sign read: "Think education is expensive. Try ignorance. Per-pupil funding: $7,000 per year. Prisoner per year: $35,000." Colorado, along with many other states, seems to have lost the desire for a good education system. Our country tends to attack public education, not support it. And that's ironic since the principle that everyone has a right to a good public education is a centerpiece of the American Dream.

Although the atmosphere was festive with bands, food trucks, and camaraderie among educators, students, and their supporters, the subject bringing us together was terribly serious. Educators were angry about the lack of funding for public education. A threshold had been reached. A prodromal moment. Around the nation, teachers in Tennessee, Oklahoma, West Virginia, Kentucky, Arizona, and now Colorado, were making the case that public education was in trouble. Teachers could not afford to live in the communities where they teach. In 2020, teachers in Colorado, who make significantly less than the national average, could go to any other state in America and have a better standard of living. There were stories of teachers having to live in their cars. There was a massive nationwide teacher shortage because of people leaving the profession. Young people were wondering why they would pursue a teaching degree to make less than someone waiting tables. In poorer school districts, they were switching to four instruction days a week instead of five to save costs. Yet, Colorado had one of the stronger and more vibrant economies in the nation. *What's wrong with this picture?*

The Colorado Education Association (CEA), which had organized the rally, was a longtime client, and I advised them on all education-related communication matters. I was there to manage the media table.

Several days before, my team and I had trained and prepared 30 speakers to talk to the media. The first thing we did was help them come up with a SOCO, a Single Overarching Communications Outcome, which became their *intention*. More than anything else, the teachers wanted to be able to help students thrive. The message wasn't so much that teachers had to pay out of pocket to close the gap left by the shortage of state funding — remember, they don't make a whole lot to begin with — but that the shortage was impacting the classroom and causing real harm to students. Schools around the state were dealing with leaky roofs, outdated textbooks, outdated or no technology, insufficient budgets for learning supplies, a lack of resources for field trips, and the pressing need to feed hungry students. Our message was about the *Students*, not the teachers. Students were our Heroes.

Then came the *training*. We coached a team of diverse spokespeople via a web-based conference program to respond to questions and make sure they were all on-message. Journalists came to the media table with specific requests. Maybe they wanted to speak to a rural educator. A Spanish language teacher. A paraprofessional. We had prepared and trained our spokespeople, our "Validator Gladiators," with a Message Triangle and the intention to always pivot back to students (The Point), no matter what the question. An important note about messaging and semantics: It was necessary to use the word, *educator*, instead of *teacher* because the former is regarded as a more inclusive term, referring to everyone involved in the overall school organization (lunchroom workers, nurses, school bus drivers, counselors, and janitors, for instance). It is a difficult word choice because national polling indicates that the word, *educator*, doesn't resonate as well with most people. *Educator* is not as clear an index as is *teacher*. But after strategizing, it was determined that it was more important to demonstrate understanding and respect for the vital role that paraprofessionals and support staff play in the educational process.

A highlight of our campaign was coordinating a segment on *NBC Nightly News with Lester Holt* that followed a teacher throughout her morning routine as she got her own kids ready for school. Overall, we placed over 150 media stories. The #RedforEd rallies dominated the local news cycle and made national headlines as part of the larger narrative that America is failing to adequately fund public education.

Our efforts began long before the events. Six months prior to the rallies, we had convened at a communications planning conference for the National Education Association in Las Vegas. As a team, we went through various scenarios and ideas. We talked about the issues: severe shortage of qualified teachers, increase in students' mental health issues, less time for teaching with more emphasis on testing. All of these issues confronting public education seemed to come down to money. Colorado was 46th in the country for per pupil funding. We zeroed in on the issue of education funding as the problem to solve. In Colorado, corporations receive billions in tax incentives to come and stay in the state. We wondered, *What if just some of that money went toward education?* Kathy Rendon, who would become the executive director at CEA, suggested the theme, "Classrooms Not Corporations." (To be honest, I was never wild about the dichotomy of 'us versus them.' Corporations want a good education system as much as anyone.

Attacking business didn't serve the larger purpose of creating a broader coalition and isolated the business community.)

We wrote out a plan using the outline grounded in Marshall Ganz's *Theory of Change* about how social movements begin and change happens.

Story of Us: We are the 38,000 members of the Colorado Education Association

Story of Now: Public education is underfunded; Students are suffering

Story of Self: Educators are underpaid and overworked; Many can't afford to live in the communities where they teach

Theory of Change: By increasing public education funding we can create better outcomes for students

We decided to do a survey of members to find out how much they spend out of their own pockets and their responses were illuminating. Over 2,500 people responded. Teachers often talk about the gap between what they *have* for their students and what they *need* — necessary resources for field trips, school supplies, copies, materials, and snacks for hungry students. Many teachers fill that gap themselves as best they can on meager salaries because they care.

From the survey we wrote the report and subsequent press release, "Teachers #OutOfPocket." In both we highlighted that our research uncovered that the average amount spent per teacher across Colorado was $656 dollars a year for the 2016/2017 school year (incidentally, teachers in rural areas spent much more than their counterparts in wealthier urban pockets, where parents contribute more). The findings startled everyone and garnered press coverage around the country. Even before the #RedforEd days, our research framed the issue in real terms that people could grasp. "Teachers#OutOfPocket" was the opening salvo of the Classrooms Not Corporations campaign we had designed in a cold sequestered conference room many months before.

The report quickly became a key resource cited in articles and discussions of why students needed more education funding.

The regularly hosted CEA Lobby Day in January, 2018, drew a slightly above average turnout of about 80 educators to the Colorado State Capitol to talk to their state legislators. But then something important happened. Other states started rumbling about striking. And Colorado educators got wind of it. They thought, *If we're actually worse off than teachers in other states, and if they're going to walk out, then why shouldn't we?* After Arizona, West Virginia, and Kentucky started walking out, the zeitgeist was swirling. Public education underfunding quickly had become one of the top issues of the day. The CEA-hosted lobby day in March brought out several hundred educators. The next one saw 2,000. The last two lobby days drew almost 5,000 the first day and 12,000 the next day. All told, about 17,000 educators had rallied.

What had started as a textbook planning session turned into a fullfledged social movement calling for more public funding for education. Spearheaded by Kathy Rendon and the communications team of Mike Wetzel and Frank Valdez, the organization planned every detail.

There is nothing like being at the center of a vortex when events and movements start to take on a life of their own. Suddenly our plan had grown into something wholly different. There were permits to secure. Sound equipment to order, bands to line up, speakers to arrange, food trucks to schedule, portable toilets to order, and all the other incidentals that go into producing an event. On our end there was a constant stream of media calls, statements, media advisories, and press releases. Hundreds of calls and requests for interviews — which made our prior training of spokespeople all the more critical.

Remember the statement from Marshall Ganz in Part 2: leaders create an environment and conditions to enable others to achieve a shared purpose. He said leaders should think, not of themselves, but of the collective whole. Such a sentiment aptly described the heart of the Days of Action. It's beautiful when excitement builds and you get a keen sense of purpose shared among your colleagues. Those precious few days and hours when it all came down to a single moment, thousands of teachers joined together voicing hope for the future and a commitment to their students, our children.

We accomplished a lot in this campaign. In the following legislative session, the *School Finance Act* increased average per-pupil investment by $475, including $150 million in the K-12 budget. This was a significant long-term funding achievement.

In his State of the State address, then-Governor John Hickenlooper laid out $30 million for attracting and retaining *rural educators*. It was a good start, but there is a lot more to do. There were a number of bills that addressed the educator shortage by seeking to expand and improve educator residency and preparation programs.

The larger point is that we had moved education to the top of the public agenda so it would get the crucial attention it needed amidst the other issues competing for funding — such as transportation, prisons, parks and agriculture — addressed in over 500 bills per legislative session. What had started in a conference room in Las Vegas a year earlier had manifested real outcomes for Colorado students.

Amie Baca-Oehlert, who served as vice-president of the Colorado Education Association at the time and became president, reflected on the #RedforEd rallies several years later. She fondly recalled the energy and excitement of those days and how proud she was to stand with her fellow educators. She told me, "I've learned just how hard it is to create and sustain a movement. It's just not that easy." The biggest lesson she drew from it all, she said, was that doing public work, such as education, defines the future you want to see.

It doesn't always come true, but if we don't define what we want public education to look like for students, then others will — for-profit education companies that run charter schools, powerful testing companies, and opinion makers who argue that we don't even need a public education system at all because everyone should be fending for themselves. It is our responsibility to provide a picture and frame and then have our movement work towards those policies, which mostly boil down to issues of funding. We have to paint the ideal classroom.

Ganz's formula worked for us in the practical application. He writes,

> A social movement tells a new story. Scholars of moral economy showed that actionable 'grievances' were experienced as an injustice, not simply an inconvenience, but as a wrong that demanded righting. Psychologists showed that grievance leads to action only if combined with efficacy, or hope.

Psychologist Jerome Bruner asserts that the act of creating a story is the mechanism by which we learn to exercise agency. We make choices about destiny in the face of uncertainty.

Our rallies and Days of Action catalyzed real change. The rally on the West Steps is the perfect illustration why communications is so important and why I love the work I do.

22

Participant, Not Spectator

Long-distance runner and 1972 U.S. Olympian Steve Prefontaine once said, "To give anything less than your best is to sacrifice the gift."

In organizations, people like to talk a lot, but rarely do they like to make decisions. Making decisions, as they see it, is a way to draw criticism and set yourself up for failure. It is easy to avoid the accountability that comes with decision-making. But if you're not making decisions, if you're not planning and executing a strategy, you're not in The Great Game.

One of the most important lessons of *Ignition* is: **Be a Participant, not a Spectator.**

There are a million things you can do with your life, but the 24-hour day forces you to make choices about what you want to accomplish and when. Over a long lunch at a French restaurant, friend and fellow swimmer Dan Richardson explained to me a program he developed that maps out how a person will most likely spend their limited time on this planet. It's called *Helix*.

Each day the sun rises and the sun sets. The earth rotates in an orbit and gives us a reference to time and space. Still, it's easy to forget that everything is in motion. Right now, for instance, the sun is spinning around the center of the Milky Way at 52,000 miles per hour. Helix gives meaning to those rotations around the sun and, based upon how you already spend your time, predicts how you will spend the rest of those spins. Those raucous (and far too few) rotations around a big ball of fire.

Dan's program takes a calendar and instead of looking at appointments, it focuses on what else you are doing. Specifically, what you are *not* doing: sleeping, taking a long bike ride, sitting on the front porch reading the Sunday *New York Times*, enjoying a long lunch, or talking to your friends. Helix looks at a cycle that is almost by definition the *anti*-calendar. By combining all sorts of data including social media profiles and emails, but mostly by simply tapping into phones, *Helix* can predict *combustion*. What happens outside your schedule? This is a very profound way of looking at your life. After all, what happens outside your appointments and organized activities might be just as important as all of those big events. As with the four-stroke engine, where there is *Ignition*, there is *Combustion*.

As *Helix* shows, our moments are brief and fleeting. Therefore, **Patience is not a virtue.** Anyone who has faced serious deadline pressure knows that patience will not get it done. There are goals and objectives to meet. Resources to nab before they run out. Votes to cast before election day. Writers have deadlines. Movies have release dates. Time is running out and deliverables are expected. There's no messing around. I hate to contradict Irma Thomas and the Rolling Stones, but in this instance, they got it wrong: Time is *not* on our side.

In communications and marketing, *immediacy* wins the day — especially when we try to manage reactions to events and crises. Right now. Let's go! *¡Vamos!* Don't wait, start thinking, start planning. Line up the necessary elements of interests and funding, create the messaging content, and train the team. Then execute. Roll with a sense of urgency. Push ahead. Venture outside your comfort zone. That slight taste of fear in the back of the throat isn't going to kill you. I go by the philosophy that if you aren't living near the edge of a nervous breakdown, then you aren't really living. How do you want to finish? With regrets? Unfulfilled potential? Or with a long sigh of contentment as you reflect on the wild ride? Patience is for people who don't have places to be, paychecks to cash, and elections to win. Believe me, deadlines are your friend.

Move with a sense of urgency. Produce. Execute. Communicate. *Now.* As the *Helix* program demonstrates, the world is moving and if you aren't creating momentum, you aren't joining those who make it go "round."

APPLICATION

Make a quick list of three areas where you can apply the Ignition Communications Planning Template at the end of Part 2. Certainly, there is a special endeavor that would benefit from your talents and energy. If it's not obvious by now, the template is not only appropriate for a "communications" plan, but as a way to approach planning for just about anything.

The way we plan and approach a communications challenge or an issue — our intention — says a lot about our character. Use communications to shape the world you want to see. Maybe it's one that is more equitable or one that is more protective of our rights and freedoms. Maybe it's one that prioritizes the protection of rivers and oceans. Maybe it's a world that values education or one that lessens our tax burden. Most of us do share the same basic values. But sometimes those values come into conflict and we end up on opposite sides, focusing on what divides us rather than on what unites us.

Resist this temptation. As the FBI's lead hostage negotiator urges, *think of the person across the table as a potential friend, collaborator.*

As storytellers, we have a responsibility to tell the truth (unless you write fiction). Those who abuse it by relying on misinformation waste a precious opportunity to contribute to the common good. Remember Edward Bernays, the founder of Public Relations, who insisted that using the power of communications for anything less than the public good is to waste the gift. As the storytellers of our time, it is our duty to seek the truth, and to bring clarity to the issues of the day.

Yet, tribal divisions are simmering to a slow boil. Most of what the world sees is the politics of hatred and division or the abdication of responsibility with a *go-it-alone approach.* History tells us that when a schism in society occurs, the factions rarely join back together. This explains why it's so important to improve our communication skills. The stakes couldn't be higher. Our ability to get along might be the ultimate test of our species. Can we, separated by tribe and ideological standpoint, unite to save our planet from disease, poverty, and slow incineration?

I happen to think we can.

Let's be clear, it's possible to master all of the techniques in this book, but the real question at this point is *How we will use the power we cultivate?* Will we harness it for the common good?

When you're on the ride, in the crisis, working for something greater than yourself, you'll find there's nothing more exciting than living with purpose and conviction. Like any organism, we are either growing and learning or we are dying. Observing is critical, but alone it is not enough. We shouldn't just look at life. We must become a part of it — not the audience, but the actors on the stage.

Upon reflection, it seems that there are remarkably few decisions that decide our fate. One fatal thing we said — or didn't say — that could have changed important outcomes in our lives, maybe even our destiny. The friends and colleagues we choose, our career, where we live, whom we marry, or if we marry at all. A phone call. Whether to write a play, join a band, or swim across the lake. These choices may seem small and inconsequential at the time, but a few of these decisions become the touchstones of life. The same goes for an organization or company: a few critical moves define the totality of the trajectory. They are our prodromal moments.

Climb your own mountain. Run. Ride. Write your own memo. Write a book. Go on an insane, glorious, painful adventure. You will fail. You will learn. Success will look different than what you thought. But remaining in that space, manifesting dreams — that is the only way anything will ever happen. Otherwise you — and the things you care about — are at the whim of someone else.

I have often thought of "the Contra Life" — the notion that your *real life*, the one you *should* be living, is going on someplace without you. One of my biggest fears is not living up to my true purpose or potential.

IGNITION PRINCIPLES

Always be Charging. Stay on the offensive.
Control the agenda
Rule of Reaction: The reaction defines the event
Explaining is losing.
Momentum: Energy begets more energy
Execute Situational Awareness:
    Understanding what you know and what you don't
Greater than yourself: Not about you;
    Intention for the bigger purpose
Participant; not a spectator
Friction: Manage it;
Story telling: The art of divination; Rise Above: Be a vessel

FIGURE 22.1
Ignition principles.

It's tough to think you aren't on your right path. Nobody understands. Watching life happen from the armchair. That sense of leaving our lives unfulfilled is the worst. Not failure, *per se,* but taking no position at all. No place at the table.

As the Native American Hopi Prophecy says, "Do not look outside of yourself for a leader. We are the ones we've been waiting for."

23

Ash Blast

Aspen Sheriff Bob Braudis's phone was blowing up. On *The Tonight Show with Jay Leno*, Johnny Depp had mentioned shooting the ashes of gonzo journalist Hunter S. Thompson out of a massive cannon in Aspen, Colorado. By the following morning, the national media started asking the author's family and the sheriff about this outrageous commemoration.

Sheriff Braudis and the Thompson family told the media: *Call Moseley.*

That's when my phone started ringing off the hook. The problem was, no one had called me to tell me what was going on. I was at a loss with reporters. *Umm, yeah, sure, let me call you back… I'm getting into an elevator. Sorry can't hear you…*

After a few calls with the Sheriff, the Thompson family, Depp's sister, Christie Dembrowski, and the Hollywood producer, John Equis, I was officially brought on to be the spokesperson and communications director for the funeral of the legendary journalist. As Depp explained on Leno, he was going to build a 157-foot tall monument, topped with "the Gonzo fist," a polydactyl (double-thumbed) fist clutching a peyote flower. The author's ashes would shoot out over the Roaring Fork Valley in a blaze of glory followed by a terrific party worthy of the author himself.

This was no accident. Decades before, in the 1970s Hunter S. Thompson had described this very funeral idea to a BBC Film crew following him around his place, Owl Farm, in Woody Creek, Colorado, outside Aspen. In the last bit of the BBC footage, Thompson scoffs, "Of course, it will be impossible to pull off without me here." *Well*, we all thought, *the joke's on you, pal.*

I was uniquely qualified and positioned for this particular role as spokesperson for Hunter's legacy. He had recruited me to support the campaign to free Lisl Auman, who at the time was serving a life sentence in prison for felony murder. Auman was unwittingly riding with a skinhead in a stolen Trans Am. High on meth, he ended up shooting a cop and then himself. The authorities accused her of handing him the gun (she didn't).

From a prison cell, she wrote Hunter a letter to tell him that his books weren't available in the prison library. He not only wrote her back, he also reported on the case and became her chief advocate. After finding out about his interest, I faxed him a memo (which he called "the Mojo Wire"). He called me soon after and thus began our national campaign to get Auman out of prison, and more broadly, to overturn the Felony Murder law.

Hunter and I produced the rally at the Colorado State Capitol, the one with Warren Zevon and a bevy of other notables and celebrities. The ensuing press propelled the case to the Colorado Supreme Court. Later, after Hunter had died by suicide, the Denver District Court set Lisl Auman free, remanding her sentence for time served and probation. Through the campaign, the Thompson family saw that I knew how to work with the media, but perhaps more importantly, they could tell that I understood the late author's legacy.

Hunter Thompson invented a style of writing that came to be called "Gonzo Journalism," in which the writer is part of the story, not merely an unbiased observer. Hunter was not a casual spectator in the events he covered; he was a character whose thoughts and language merged with the narrator's perspective. Many believe *Gonzo* means something about booze, guns, and acid, but that's not it. The very essence of Gonzo Journalism is about becoming part of the story. Not just watching and observing, but participating.

After working out the details of my engagement with Depp, I spent the next six weeks at a Production House outside of Aspen with a constantly blaring radio and all the ensuing hoopla and drama. It was our staging area for every aspect of the event. Our first job was to hire around-the-clock security. We were going to need it with all the media attention already on the Ash Blast.

While the run-up to the funeral required a lot of heavy lifting (literally), for me, it was essentially a communications exercise. Here I was, in cowboy boots and a seersucker suit, the voice of a multimillion-dollar psychedelic, pyrotechnic funeral that would include a cast of freaks, senators, bohemians, and oddballs... in addition to the more established deadbeats. I knew from taking on such heavy responsibility before that these kinds of assignments carry a certain breathless thrill, but also heavy tension and anxiety. Friction. Being a spokesperson in these situations requires hypervigilance. There is only one chance to get it right — at least, in the case of a funeral.

My first step was to work with the family to design the invitation, which was a simple white, super heavy card stock with a silver embossed Gonzo dagger. A quote on the invite from Hunter read,

My main luxury in those years — a necessary luxury, in fact — was the ability to work in and out of my home-base fortress in Woody Creek. It was a very important psychic anchor for me, a crucial grounding point where I knew I had love, friends, & good neighbors. It was like my personal Lighthouse that I could see from anywhere in the world — no matter where I was, or how weird & crazy I got everything would be okay if I could just make it home. When I made that hairpin turn up the hill on Woody Creek Road, I knew I was safe.

FIGHTING FOR CONTROL OF THE NARRATIVE

If there is anyone who understood the alchemy of words it was Hunter S. Thompson. He penned groundbreaking books, including *Fear and Loathing in Las Vegas*, which was made into a Hollywood movie featuring Depp (who also starred in another Hunter adaptation, *The Rum Diary*). In the 60s, Hunter embedded himself with the notorious biker gang to write *Hell's Angels: A Strange and Terrible Saga*. He chronicled the Richard Nixon and George McGovern presidential race in *Fear and Loathing on the Campaign Trail '72*. My sister, Mary, gave me a copy of this book when I was 18 years old and it inspired me to pursue a career in politics and, ultimately, communications. The way Hunter told it, the campaign trail could be a hell of a ride. The urgent storytelling portrayed politics as an adventure that actually meant something. As I reflect years later, you might even say the funeral was a point in my own Hero Cycle and that the blasting of the ashes was a sort of reconciliation with a Father Figure.

My job as the funeral spokesperson should have been relatively easy. From the earliest planning stages the policy was: NO MEDIA ALLOWED IN THE EVENT. No *ifs, ands,* or *buts*, according to Depp, who was the head honcho since he was footing the bill. But that still didn't stop hundreds of reporters and media trucks from infiltrating Aspen. With them came the Gonzo Pilgrims. Untold numbers of people who, like their hero, wanted to be part of the action. A class of literature students from New Orleans sat on the hillside watching. Hunter impersonators roamed Woody Creek. A friend of mine ran a small security operation for a Saudi Prince who watched the proceedings from a high-powered military telescope on the mountaintop. The whole scene was devolving into complete madness.

A valuable lesson in Ignition: if you don't allow the media to talk to your sources, they will consult others who may have entirely different interests and be all too happy to fill the information vacuum. In this case, reporters asked the locals how they felt about being left out while all these Hollywood-types were invading Aspen.

To hell with 'em!

A narrative began to develop, *This is exactly the kind of exclusive Hollywood bullshit Hunter would have loathed.*

We began to lose control of our own story.

One person who called me up looking for an invitation said something like, "I was Hunter's drug dealer for twenty years! What do you mean I can't get in?"

Hunter's benefactor and longtime friend from down the road, George Stranahan, the physicist and whisky maker, had trouble getting his own family into the funeral. Wayne Ewing, Hunter's documentarian and good friend, was told to leave his camera at home. Aspen wags were incensed at the spectacle. Truth be told, although Johnny Depp was spending $3.5 million, there wasn't even enough money to pay Owl Farm's electric bill. There was a kernel of truth to the counter-narrative that could not be ignored. I was in a difficult spot.

Katharine Q. Seelye from the *New York Times* flew into Aspen to report on the funeral. She had covered Al Gore's presidential campaign, and we were acquainted with each other. This time, I had to deliver the unfortunate news that she was not allowed at the big event. She would join a chorus of journalists who were badmouthing me in the J-Bar at the Hotel Jerome late at night. I got word they were grumbling about the lack of information and access. *So unprofessional.*

The next day, I received a call from an editor at *The Times*. He was enraged. *Who the hell do you think you are? We are The New York Times and we have to be there.*

I told the editor to hold on a minute — *The Times* wasn't allowed into Ernest Hemingway's funeral either. Losing my patience, I jabbed, "You watched it from outside the fence!"

Then I thought of something. I mentioned that Hunter had always wanted to be on the front page of *The New York Times*. And then I went for the brass ring. I said, "We'll consider letting you in… if you make Hunter A1." Front Page. Preferably above the fold. The editor became defensive, paused, and grumbled something noncommittal. *I don't know. There's a lot happening in the world…*

I responded, "Yeah, and there's a lot happening right here in this little valley."

The editor wouldn't make any guarantees. And neither could I about access for Seelye.

Douglas Brinkley, history professor at Rice University, commentator for CNN, and Hunter's literary executor, was my guiding hand through the battlefield. He and I conferred and agreed that something had to be done. We needed to be more proactive if we wanted to take back control of the narrative. Remember the advice from Part 1: *Silence can be Deadly.*

Brinkley also gave me a charge: This wasn't just an explosion or a party, it was a chance, with all the media attention, to frame the author's unique legacy and contribution to American letters. It was an opportunity to tell the story of one of the most interesting and original writers of the second half of the century. We strategized about the messaging — it needed to be about a writer who had changed the face of journalism. The party was just a hook.

HELICOPTER WARNING

To try to reclaim the narrative, I convinced Depp and the Thompson family to let me organize a media tour of the property for a pool of reporters from the largest news bureaus. After much consternation and hand-wringing, they agreed. I invited about a dozen journalists, including Seelye and reporters and photographers from the *Associated Press*, *Reuters*, and state and local papers. The tour generated mounds of positive press and gave people an inside look into the construction of the Gonzo Monument. Their expensive trip to Aspen was not wasted. I had trained a few people from the assembly crew to explain the intricacies of building the 157-foot tower. We had pulled back the curtain and the sneak peek we offered was doing the trick. We were regaining some measure of control.

Afterward, Brinkley and I went out on a limb and made a side arrangement with Seelye. She could attend the event in the guise of a staffer for the former presidential candidate and subject of Hunter's book on the 1972 election, Senator George McGovern, who would be attending. She couldn't bring a notepad, camera, or recording device. And she had to leave directly after the ceremony and before the celebration. I escorted her out to the arranged driver standing by for her exit. Not even Depp knew about our scheme. Douglas Brinkley whispered to me, "We're on Legacy Patrol."

Still, the chaos continued. Paparazzi were swarming like turkey buzzards. Jimmy Ibbotson of the Nitty Gritty Dirt band fired a shotgun at a photographer snapping pictures from the bushes. I had to issue a helicopter warning because one had nearly grazed one of the guy-wires securing the structure to the ground while trying to get close-ups. At one point I hurled my phone against the wall in exasperation.

One of my favorite press releases was to report that we had cleared the airspace over Woody Creek. With media buzzing in the air and on the ground, we worked with the Federal Aviation Administration to restore some order. Naturally, the helicopter warning advisory to the media only served to heighten the drama and suspense.

Lesson: If you want to attract more media to a funeral, announce that the FAA has cleared the airspace.

A FULL MOON RISING

The Ash Blast, as my wife coined it, was utterly subversive. Surreal. Raw. Explosive. Ridiculous. *Expensive.* The drama of humanity itself wrapped in a silver Gonzo dagger.

There were musicians Lyle Lovett and David Amram, senators, painters, polo players, actresses, an accused murderer, producers, a sheriff, Pulitzer Prize winners, publishers, criminals, and a shit-ton of lawyers. Unique spirits of the universe had congregated on this surreal night in a mountain meadow to say goodbye.

Buy the ticket. Take the ride, Hunter commanded.

Before long, Bill Murray was waltzing across the floor with a blow-up sex doll. Jann Wenner from *Rolling Stone* was grinning ear to ear, high on psilocybin mushroom chocolates in the shape of a Gonzo Fist. The *Denver Post* reported that Lynn Goldstein, the rock and roll photographer who toured with the Rolling Stones, had to be taken out by stretcher.

The event itself was divided into two parts. The first part was ceremonial. Everything, including the monument, was draped in black or red cloth. There were a few remarks, a Japanese drum band slowly built to a crescendo as the shroud — stretching higher than the Statue of Liberty — was slowly sucked up into the belly of the monument for the Grand Unveiling. Kudos to John Equis, the stage master and producer. Equis, who had helmed numerous Oscars productions among other Hollywood events, was well-practiced in the Art of the Reveal.

The monument was custom-designed and handcrafted in Los Angeles by the company Design Setters. Four "Oversized Load" 18-wheel trucks drove the finished pieces to Aspen for assembly. (The whole building of the monument and event are chronicled in the excellent documentary by Wayne Ewing called *When I Die.*) Anita Thompson, Hunter's widow, had flown out with his ashes to the famed Zambelli Fireworks Company in New Castle, Pennsylvania, where they packed the ashes in canisters with fireworks and "Starblasts." Matt Wood, the designer and producer from Zambelli fireworks, warned her that the explosion would be just below the level of a sonic boom (which, incidentally, required me to issue a Small Pet Advisory to Woody Creek residents). More press. More attention. More Drama.

The big moment arrived. Cued to Norman Greenbaum's 1969 anthem, "Spirit in the Sky," the first fireworks screamed into the night, illuminating upward-gazing faces. A minute of black sky followed building suspense. We grew more and more impatient every second. Finally, toward the tail-end of the song, Hunter's ashes were blasted out.

Boom. People.

It wasn't just a funeral — it was a celebration of a literary style and an entire approach to looking at the world. The funeral had to embody the contradictions of the man himself. A saint and a sinner. An author who was one of the first to put himself into the story in a modern way, elevating journalism beyond the he-said/she-said dichotomy.

As the sun started to rise, my wife, Kristin, and I, along with Steve Cohn, who had supervised the building of the monument, his running partner, Mako, who was the former manager of the Viper Room nightclub in Los

Angeles, my friend and attorney, Tom Ward, his wife, Dru Nielsen, who arranged the plea deal for Lisl Auman to get out of jail, the future mayor, Torre, and my cousin, Glynde Mangun (aka Mango), all piled into the Red Shark, Hunter's replica of the convertible he drove on the fateful trip that inspired *Fear and Loathing in Las Vegas*. Behind us the dawning light silhouetted the massive Gonzo Monument against the mountains of the Roaring Fork Valley.

"We're on the road to nowhere!" someone shouted. We howled with laughter and gave cheers and hugs. Exhausted after weeks of intense, around-the-clock work, we marveled at the sheer magnitude of what had just transpired.

These are moments when the deepest friendships are forged.

My quote to BBC Radio that night was, "With a full moon rising over Woody Creek, there was no finer place to be on the entire planet than when Hunter S. Thompson's ashes were shot out into the ether."

THE REVEAL

The New York Times piece came out two days after the funeral, with Hunter on the front page as a teaser. The story filled most of page A4. Seelye wrote,

> At the entry to what could only be called the set, his portrait was hung at the center of his personal literary solar system, surrounded by the planets of Samuel T. Coleridge, Joseph Conrad, William Faulkner, F. Scott Fitzgerald, Ernest Hemingway, Henry Miller, John Steinbeck and Mark Twain.

This single paragraph in the *New York Times* put Hunter smack dab in the bullseye of the literary pantheon.

He was famous before, but now he was a legend.

Seelye wrote, "By nightfall, scores of fans had gathered at the nearby Woody Creek Tavern and outside the gate to the property. Sheriff's deputies said that 'numerous people' tried to crash the scene but were escorted away."

She reported on the BBC documentary in which Hunter described in detail how he wanted his ashes dispersed. She wrote of how

> The silky red dressing around the monument slowly unpeeled itself, revealing a rocket-like structure embedded with a dagger. It was crowned by

Mr. Thompson's logo, a two-and-a-half-ton red fist with two thumbs and a psychedelic peyote button pulsating at its center, a Day-Glo sight visible for miles around.

I especially appreciated this last part — that **The Reveal**, which was no small effort, made it into *The Times*. This is another valuable lesson I took from the funeral. A key tactic of getting people to care involves the choreography with which a concept is introduced. First impressions mean everything. As Tommy G. understood when he created the Butterfly, *ooohhs* and *aaahhs* are the tinder of caring.

The firing of the ashes from the gonzo monument was not just a metaphor for the concept of *Ignition* illustrated throughout this book; in reality, it was an act of Ignition in and of itself. That's Ignition with a capital "I." As we learned in fluid mechanics, *Ignition* creates *Power*.

The Ash Blast wasn't just a funeral. This was an exercise in getting people to care. How do you inspire people to think once more about Hunter Thompson as an author and elevate their appreciation of his unique contribution to American literature? *Blow his ashes out of a cannon.* Every literature student who researches Hunter will probably read those lines of Seelye's in *The New York Times*. But to reserve Hunter's rightful place in the pantheon of American writers, we didn't have to say anything at all. Blowing him out of a 157-foot tall fist made The Point more eloquently than any words ever could have.

Res ipsa loquitur.

24

Your Rock — Your World

For his passion for life and disregard for their constructs, the gods cast Sisyphus into the pit where every day he was forced to roll a rock up a hill. He was there for eternity at his never-ending task, but he wouldn't be alone. He'd have his rock. His only companion. His single purpose.

One can imagine his pained face pressed tight against the rock, strained muscles bracing against the forces of gravity. At the end of the long toil, as he reaches the apotheosis, which knows no limits of space or time, he watches helplessly as the rock rolls back down. It descends into a world of doubt. Of chaos and uncertainty. Only in pushing and pushing, does Sisyphus find meaning. The cycle of *Destruction and Creation*, as Colonel John Boyd wrote.

Sisyphus walks down to meet his destiny at the bottom of the canyon. In those moments when he trods down, he is aware. He is observing. As in the four-stroke engine, Sisyphus is taking in fuel, motivation. He is readying himself for Action and Energy. He is Humanity itself, reflecting on history and dreaming of the future. In this space where he considers his fate (orienting), he has cultivated a perfect awareness of himself and his situation. As he descends to his rock, he is not despondent. He is fully aware and perhaps somewhere down deep he knows that he is still superior to a fate defined by endless punishment.

We are Sisyphus. At the bottom of the hill, we conceive of another intention. Another purpose. Another adventure. Sisyphus possesses purpose and conviction as he starts his journey anew and starts to push the rock uphill once more. In this moment, Sisyphus is stronger than his rock. He raises a giant middle finger to the gods. This torture of pushing a rock is also his redemption. When we prevail personally, we also prevail collectively. We endure as a species.

I could not have fathomed as that high school junior at the National Debate Championships in Chicago that the story of Sisyphus would stay with me throughout my life. I often think of the metaphor of The Rock when I need to summon the motivation to get up in the morning to swim or to muster the courage and energy to get to work. To write. Sometimes I think of The Rock in coping with what occasionally feels like "the tedium of everyday life." By now, The Rock isn't just a metaphor for me; it's a way to live life, a perspective on the world. Like a center of gravity, it pulls my locus of control within.

There are three valuable lessons here. The first is that communicating itself is a rock that we must continually push. Every day, from the moment we awake, we are pushing. And every day it rolls down upon us. As such, our task is never done. This Rock is never still. Each day we must prove ourselves once more by keeping it in motion. Just keep pushing. Keep swimming. Keep writing. Another issue. Another crisis. Another rock to push up the hill.

When any organism ceases to communicate, it starts to die. Like Sisyphus, if we stop pushing, we're done. No matter what you do — medicine, law, education, if you run a startup or a nonprofit — if you're an engaged person who has ideas for making the world a better place and you get up every morning and get right to it, your work is never done. You are always pushing. If you're not, then you're fading. Flailing. Failing at your true purpose.

The second lesson is even more important. Sisyphus has his rock. *What is your rock? What are you pushing to the edge?* I spend my life pushing rocks. Every day. For my clients, for my children and wife, for my friends, and for myself. As with Sisyphus, it's never over.

And the third lesson is this: when we develop our own awareness, we can transcend the monotony of daily life. We all *need* a rock. Without our work, there is no purpose. Like Sisyphus, we should rely on our work, our rock, to rise above the absurdity of daily life and gain clarity and inner peace. In a way, cultivating that greater consciousness constitutes our *greatest* work.

When we say *Yes*, when we own our rock, we commit to make our efforts endless and unceasing. They will be hard, they will be grueling, but in the end, they will be gratifying. As long as we keep that in mind, as long as we keep pushing, we become the master of our own universe.

We control our environment.

MONADO GUMBO

In New Orleans, in religion, as in food or race or music, you can't separate nothing from nothing. Everything mingles each into the other — Catholic saint worship with gris-gris spirits, evangelical tent meetings with spiritual-church ceremonies — until nothing is purely itself but becomes part of one fonky gumbo.

Mac Rebenack, Dr. John, The Night Tripper

UNDER A HOODOO MOON

To depict the crystallization of a thought, or the distillation of all possibilities into one worthy idea, my daughter Amelia and I created the *Monado Symbol*. The many into the one.

ABOUT THE MONADO SYMBOL

Another wonderful way to illustrate this process is through the Monado Gumbo.

The MONADO GUMBO is something richer, spicier, more delicious than the sum of its parts. I might send out a call to friends 24 hours in advance:

"Make Gumbo Not War. Come by tomorrow if you're available." The next afternoon I assemble simple ingredients:

Flour, oil, onion, green pepper, celery, green onion, a chicken, andouille sausage, spices. At first, it's chaos on the kitchen counter.

The top lines of our Monado.

First start the **roux**. Equal parts flour and oil. Maybe a little bacon grease or butter. The traditional method is to cook it in a skillet, but Ms. Jenny Blow taught me how to bake it in an oven. Slow heat at 325, stirring every once in a while. I measure this by stirring every time I make a cocktail. Speaking of, it may not be a bad time to call your mama or your daddy, maybe Cousin Mango. Say hello. Then get ready for the journey. Good music should be playing: Kermit Ruffins, Galactic, Diplo, Trombone Shorty, Allen Toussaint, Dr. John, or some other funky stuff.

Throw a whole chicken into a big pot of boiling water and let that just cook away while you do everything else.

Next comes the fun part. Prepare the **mirepoix**. In South Louisiana they call it the Holy Trinity of Creole cooking: Yellow onions, green peppers, and celery, all chopped and cooked together until clear. Those three elements cook to form a unified substance all its own. This is the base for almost every creole dish and provides the essential taste of the deep bayou. Gumbo, Sauce piquante, stuffed redfish, crabmeat dressing — just about everything starts with The Holy Trinity. This is when I like to call a good friend or my sister, Mary. Hey, what's happening? Friends are an important part of life, let them know you care. Make them a part of the experience of cooking. The gumbo starts to take on love as the kitchen fills with the aromas of the mirepoix and the chicken, and the music puts a spring in your step.

As the mirepoix is cooking, render the sausage in a sauté pan. Then debone the chicken while preserving the stock.

Combine the mirepoix with the stock in a big pot. Stir it all around. Do the hokey pokey. Then add the roux. Watch out, this is hot boiling grease hitting a boiling pot of water. Get ready for splash and sizzle. Pour slowly and cautiously, protecting your hands. Next throw in all the meat. Stir it all together.

Now turn up the music. Make another cocktail. Call another friend. Stir the pot. The gumbo becomes one as it simmers for a few hours.

Later that evening, friends start showing up. People eat from coffee mugs filled with a little brown rice and gumbo with a slice of french bread. They break bread together and the heat from the gumbo brings them closer in the shared experience. Old friends and new ones. The crystallization of all the cooking and work assembling the ingredients and the people.

More than just flour and oil. More than just onions, peppers and celery. This is not just soup. It's something greater than the sum of its parts. Eating gumbo is a defining moment for all involved, cooks and guests. Magic can be defined in many ways, but this is a simple, fitting description.

References

PART 1

Alinsky, Saul. "Rules for Radicals". *Random House*, 1971.

Ambjørn, Jan, et al. "The Self-Organizing Quantum Universe". *Scientific American*, 2008.

Bennett, Jeffrey. "Max Goes to the Moon". *Big Kid Science*, 2003.

Bennett, Jeffrey. *"What is Relativity: An Intuitive Introduction to Einstein's Ideas, and Why They Matter"*. Columbia University Press, 2016.

Bernays, Edward. *"Crystallizing Public Opinion"*. Boni & Liveright: New York, 1923.

Bhasin, Kim. "9 PR Fiascos That Were Handled Brilliantly by Management". *Business Insider*, 26 May 2011.

Byerly, Carol R. *"Fever of War: The Influenza Epidemic in the U.S. Army during World War I"*. NYU Press, 2005.

Chrisomalis, Stephen. "The Egyptian Origin of the Greek Alphabetic Numerals". *Antiquity*, vol. 77, no. 297, 14 October 2002, pp. 485–496. doi:10.1017/s0003598x00092541.

Cuddy, Amy. "Your Body Language Shapes Who You Are". *TED Talk*, June 2012.

Eccles, Robert G, Newquist, Scott C, and Schatz, Roland. "Reputation and Its Risks". *Harvard Business Review*, February 2007.

"The Engineering of Consent". *Annals of the American Academy of Political and Social Sciences*, March 1947.

Fink, Steven. "Crisis Management: Planning for the Inevitable". *iUniverse*, 2002.

Frankl, Viktor. *"Man's Search for Meaning"*. Translated by Ilse Lasch. Beacon Press, 2006.

Freedman, Lawrence. *"Strategy: A History"*. Oxford University Press, 2013.

Fritz, Robert. "Path of Least Resistance: Learning to Become the Creative Force in Your Own Life". *Fawcett Columbine*, 1989.

Gage, Beverly. "'Strategy' May Be More Useful to Pawns Than to Kings". *The New York Times Magazine*, 3 September 2018.

Ganz, Marshall. *"Leading Change"*. Harvard Business Press, 2010.

Garcia, Helio Fred. *"The Agony of Decision: Mental Readiness and Leadership in a Crisis"*. Logos Institute for Crisis Management and Executive Leadership Press, 2017.

Giardetti, J. Roland, and Oller, John. "Images That Work: Creating Successful Messages in Marketing and High Stakes Communications". *Quorum*, 1999.

Harari, Yuval. *"Sapiens"*. HarperCollins Publishers, 2015.

Joelson, Richard B. "Locus of Control". *Psychology Today*, 2 August 2017.

Lasswell, Harold D. "The Theory of Political Propaganda". *The American Political Science Review*, vol. 21, no. 3, 1927.

Lippmann, Walter. *"Public Opinion"*. Harcourt, Brace, & Co., 1922.

Osgood, Charles E, et al. *"The Measurement of Meaning"*. University of Illinois Press, 1957.

Parks, Shoshi. "How Deaf Children in Nicaragua Created a New Language". *Atlas Obscura*, 13 July 2018.

Rathje, Steve. "Why People Ignore Facts". *Psychology Today*, Sussex Publishers, 25 October 2018, www.psychologytoday.com/us/blog/words-matter/201810/why-people-ignore-facts.

Richards, Chet W. "Certain to Win: The Strategy of John Boyd, Applied to Business". *Xlibris*, 2004.

Rosengren, Karl E. *"Communication: An Introduction"*. Sage Press, 2000.

Sivers, Derek. "First Follower: Leadership Lessons from Dancing Guy". *Youtube*, 11 February 2010, www.youtube.com/watch?v=fW8amMCVAJQ&ab_channel=DerekSivers.

Thomas, William I, and Thomas, Dorothy Swaine. *"The Child in America: Behavior Problems and Programs"*. Alfred A. Knopf, 1928.

Voss, Chris. *"Never Split the Difference"*. Harper Business, 2016.

PART 2

Anderson, Monica, and Jiang, Jingjing. "Teens, Social Media & Technology 2018". *Pew Research Center*, 31 May 2018.

Aristotle. The Poetics.

Campbell, Joseph. *"The Hero with a Thousand Faces"*. Pantheon Books, 1949.

Friedman, Vanessa, and Bromwich, Jonah E. "Cambridge Analytica Used Fashion Tastes to Identify Right-Wing Voters". *The New York Times*, 29 November 2018.

Gamhegawe, Gaya. *"Effective Communications Participant Handbook: For WHO Staff"*. WHO Press, 2015.

Gladwell, Malcolm. *"The Tipping Point"*. Little, Brown and Company, 2000.

Grassegger, Hannes, and Krogerus, Mikael. "The Data That Turned the World Upside Down". *VICE*, 28 January 2017.

Lakoff, George, and Johnson, Mark. *"Metaphors We Live By"*. University of Chicago Press, 1980.

Lenhart, Amanda. "Teens, Social Media & Technology Overview 2015". *Pew Research Center*, 9 April 2015.

Mehdi, Yusuf. "It's Better with the Butterfly: MSN 8 Offers Advanced Communication Tools, Powerful Browsing and Enhanced Security Features". *Microsoft, Microsoft*, 14 October 2002.

Naisbitt, John. *"Global Paradox: The Bigger the World Economy, the More Powerful Its Smallest Players"*. William Morrow & Company, 1994.

"Red Brain, Blue Brain". *Hidden Brain from NPR*, 8 October 2016, https://www.npr.org/transcripts/654127241.

Resnick, Brian. "22 Percent of Millennials Say They Have 'No Friends'". *Vox*, 1 Aug 2019.

Westen, Drew. *"The Political Brain: The Role of Emotion in Deciding the Fate of the Nation"*. PublicAffairs, 2007.

PART 3

Davisson, Megan. "State Budget Briefing FY2018–19: Department of Corrections". *Joint Budget Committee of the State of Colorado*, 20 December 2017.

Dickhoner, Brenda Bautsch, and Brown, Chris. "Dollars and Data: A Look at K-12 Education Funding in Colorado". *Common Sense Institute*, 28 August 2019.

Lehrer, Jonah. "The Science of Irrationality". *WIRED*, 18 October 2011.
Seelye, Katharine Q. "Ashes-to-Fireworks Send-Off for an 'Outlaw' Writer". *The New York Times*, 22 August 2005.
Tversky, Amos, and Kahneman, Daniel. "Rational Choice and the Framing of Decisions". *The Journal of Business, Political Science*, vol. 59, no. 4, Pt 2, October 1986.

ADDITIONAL REFERENCES

Anderson, Walt. "*The Truth About the Truth: De-Confusing and Re-Constructing the Postmodern World*". G. P. Putnam's Sons, 1995.
Ansolabehere, Stephen, and Iyengar, Shanto. "*Going Negative: How Attack Ads Shrink and Polarize the Electorate*". Free Press, 1996.
Aristotle, and Barker, Ernest. "*The Politics of Aristotle*". Oxford University Press, 1962.
Benoit, William L. "*Communication in Political Campaigns*". Peter Lang, 2007.
Blundell, William E. "*The Art and Craft of Feature Writing: Based on the Wall Street Journal Guide*". *Plume*, 1988.
Chrislip, David D, and Larson, Carl E. "*Collaborative Leadership: How Citizens and Civic Leaders Can Make a Difference*". Jossey-Bass, 1994.
Cicerón, Marco Tulio, et al. "*Cicero: On Duties*". Cambridge University Press, 1991.
Clapham, Christopher. "*Third World Politics: An Introduction*". Croom Helm Ltd, 1985.
Collins, Jim. "*Good to Great: Why Some Companies Make the Leap... and Others Don't*". HarperCollins Publishers, 2001.
Emoto, Masaru. "*The Hidden Messages in Water*". Atria Books, 2005.
Gallo, Carmine. "*The Storyteller's Secret: From TED Speakers to Business Legends, Why Some Ideas Catch on and Others Don't*". St. Martin's Griffin, 2017.
Gardner, Howard, and Laskin, Emma. "*Leading Minds: An Anatomy of Leadership*". Basic Books, 1995.
Gecan, Michael. "*Going Public*". Beacon Press, 2002.
Gerzon, Mark. "*Leaders beyond Borders: How to Live - and Lead - in Times of Conflict*". Mark Gerzon, 2003.
Jiménez, Francisco. "*Poverty and Social Justice: Critical Perspectives: A Pilgrimage toward Our Own Humanity*". Bilingual Press/Editorial Bilingüe, 1987.
Katzmann, Robert A. "*Institutional Disability: The Saga of Transportation Policy for the Disabled*". The Brookings Institution, 1986.
Kemmis, Daniel. "*Community and the Politics of Place*". University of Oklahoma Press, 1992.
Levine, Michael. "*Guerrilla Pr Wired: Waging a Successful Publicity Campaign Online, Offline and Everywhere in Between*". McGraw-Hill, 2003.
Machiavelli, Niccolò, and Machiavelli, Niccolò. "*The Prince*". University of Chicago Press, 1985.
McCool, Daniel C. "*Public Policy Theories, Models, and Concepts: An Anthology*". Prentice Hall, 1995.
Ogilvy, David. "*Ogilvy on Advertising*". Random House, 1985.
Oller, John W, and Giardetti, J Roland. "*Images That Work: Creating Successful Messages in Marketing and High Stakes Communication*". *Quorum*, 1999.
Peretti, Chelsea, et al. "*The Gawker Guide to Conquering All Media*". Atria Books, 2007.

Plato, and Francis, Macdonald. *"The Republic of Plato"*. Oxford University Press, 1945.

Plato. *"Dialogues of Plato"*. Washington Square Press, 1950.

Plato, et al. *"The Last Days of Socrates: Euthyphro, the Apology, Crito, Phaedo"*. Penguin Books, 1954.

Popkin, Richard H. *"The Philosophy of the 16th and 17th Centuries; Edited and with an Introduction by Richard H. Popkin"*. Collier-Macmillan, 1966.

Pressman, Jeffrey L, and Wildavsky, Aaron. *"Implementation: How Great Expectations in Washington Are Dashed in Oakland"*. University of California Press, 1984.

Sophocles. *"The Three Theban Plays: Antigone, Oedipus the King, Oedipus at Colonus"*. Penguin Books, 1984.

Stauber, John C, and Rampton, Sheldon. *"Toxic Sludge Is Good for You: Lies, Damn Lies and the Public Relations Industry"*. Common Courage Press, 2002.

Tocqueville, Alexis de, and Mayer, JP. *"Democracy in America"*. Harper & Row, 1988.

Watson, Brian. *"The 7 Rings: A Journey to a Balanced Life of Peace, Passion, and Purpose"*. Brian Watson, 2016.

Index

Note: *Italic* page numbers refer to figures.

Milton Keynes UK
Ingram Content Group UK Ltd.
UKHW051017071024
449327UK00017B/447